Sensors and
Transducers

Sensors and Transducers

Keith Brindley

Heinemann Professional Publishing

Heinemann Professional Publishing
22 Bedford Square, London WC1B 3HH

LONDON MELBOURNE AUCKLAND

First published 1988

British Library Cataloguing in Publication Data
Brindley, Keith
　　Sensors and transducers.
　　1. Transducers
　　I. Title
　　621.37'9　　TJ223.T7

Printed in Great Britain by
Redwood Burn Ltd, Trowbridge

ISBN 0 434 90181 4

Contents

Preface

There is a considerable number of transducers currently available. Look through any electronic component distributor's catalogue and you will find a wide variety of types, and each type has many versions. On the face of it, it is not easy to choose a transducer correctly for a particular function: in many transducer specifications, terms and procedures are referred to which might deter an engineer from using a transducer when, in fact, that transducer may be the best to suit the task. Yet, opting to use a transducer merely because it is easier to interface into the measuring system is not the answer. A greater knowledge of all types of transducer capable of doing the task is the ideal, and only then can a totally satisfactory decision be made to use one transducer.

In a nutshell, this is the main aim of my work here. I have tried to bring together as much of the required information as possible in this book, so that engineers can view transducers in perspective and then make an informed choice. In no way have I tried to make an encyclopedic volume. Nevertheless, I hope enough information is given to cover most, if not all, types of sensing transducers.

Aware of the fact that transducers are known by different names in different technical disciplines, I have also tried to rationalize the varied names into categories which can be acceptable to anyone who encounters transducers. That is not to say, of course, that any particular name is wrong (after all, who would prefer to call a thermistor a *temperature-sensing semiconductor resistive transduction device with a negative temperature coefficient of resistance?* Certainly not me, just that categorized definitions of transducers make it easier to locate a particular transducer for a particular task.

The categories I have chosen to split transducers into are by measurand, i.e., the physical quantity which is to be measured. However, aware of the fact that a simple measure of a particular measurand quantity can sometimes be electronically adapted to give the measure of a different, yet related, measurand (for example, speed can be determined by measuring the distance covered and dividing it by the time taken), I have chosen to give readers the option of referring to transducers capable of measuring other related measurands where appropriate.

Keith Brindley

1
Introduction

The role of the transducer

In the strict sense of the word, defined by *The Oxford Dictionary*, a **transducer** is a device which converts variations of one quantity into those of another. In terms of electronics, however, a transducer is usually taken to be a device that converts a non-electrical physical quantity (known as the **measurand**) into an *electrical* signal, or vice versa. There are, of course, exceptions to this rule.

It follows that a transducer would be used in an electronic system to present the equipment with an electrical signal which represents the measurand under observation. Alternatively, a transducer would be used at the output of a system to generate, say, a mechanical movement in response to an electrical control signal. A basic example of transducers is a public address system; in which a microphone (the input transducer) converts sound (the measurand) into an electrical signal which is amplified than applied to a loudspeaker (the output transducer), producing sound very much louder than that picked up by the microphone.

Often the measurand, as its name suggests, is simply being measured by the electronic system and the result merely displayed or recorded, but in some cases the measurement forms an input to a controlling circuit which attempts to either maintain the level of the measurand to a predetermined level, or control a variable quantity in accordance with the measurand. Although there is considerable overlap between the use of transducers in these two examples, there is an understandable trend by the participating engineers to class the specific disciplines involved (**instrumentation** and **control**) as separate. Further, within the two disciplines there are

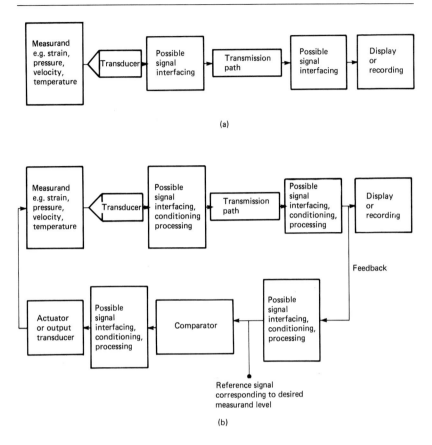

(a)

(b)

Figure 1.1 *Block diagrams of instrumentation and control systems. (a) a typical instrumentation system; (b) a typical control system. Control systems may often be considered to comprise an instrumentation system plus extra control circuits*

subdisciplines such as telemetrical instrumentation (in which measurement systems use a radio link between transducer and display equipment), chemical analysis (in which the systems used determine and display the relative abundance of constituents of a mixture), or process control (in which an industrial process, say, a steel rolling-mill, is monitored and controlled) and many, many more.

Figure 1.1(a) illustrates the main components of a typical instrumentation system; not all of these may be necessary, however, in a particular application. Figure 1.1 (b) illustrates, similarly, a typical control system. In essence, part of the control system is an instrumentation system. So, for the purposes of this multidisciplinary book, the transducers involved and the required interfacing circuits will be looked upon as common, and few further references to individual disciplines will be made.

The main points of note in Figure 1.1 are the following:

- The **measurand**; the physical quantity to be measured, such as acceleration, displacement, force, flow, level, position, pressure, strain, temperature, velocity etc. In some instances the measurand itself may be an electrical quantity such as current, voltage or frequency, and is converted to an electrical signal acceptable to the remaining part of the system – the transducer is merely an electrical conversion element.
- An **input transducer**, converting the measurand into an electrical signal acceptable to the remaining part of the system. In truth, although most input transducers do generate an electrical output, some generate other forms of signals – for example, pneumatic pressure – such input transducers are relatively few, and these are purposefully ignored here. However, transducers exist which use such non-electric output transducers merely as a *sensing element*; other parts of the transducer then convert the non-electric signal into an electrical form. All transducers function by analogue principles so, in all but a few instances, the signals produced by transducers are consequently of analogue form.
- Some sort of **transmission path**, between the input transducer and rest of the system. This may be nonexistent in terms of system operation, say, if the input transducer is located within centimetres of the rest of the system; alternatively, if the input transducer is located any distance from the system, steps must be taken to ensure the transmission path has little or, preferably, no effect on the operation of the system.
- Where a significant transmission path is involved it is sometimes necessary to provide one or more **signal interfacing** stages, to condition the small input trans-

ducer output signal by amplification, analogue-to-digital conversion, filtering, modulation etc., so that information present in the input transducer signal is not lost in its transmission to the rest of the system. These stages may include a level of data processing, where the data contained in the signal from the input transducer is digitally processed so that the resultant signal or computation may be displayed, recorded, or used for control purposes. Signal interfacing stages may occur at a number of points within the system.

It is often difficult to say just exactly where in a system the analogue transducer signal becomes data; consequently it is often impossible to distinguish between analogue signal conditioning and data processing stages. Fortunately, the distinction is rarely important.

- A **display** or **recording** device, which displays the current value of measurand for the convenience of a system operator, or which records the pertinent information for later operator use.
- In the case of a control system (Figure 1.1(b)) some form of **comparator** device is used to compare the processed data with some desired **reference** value, producing a **difference signal**.
- An output transducer, operated by the difference signal, which is used to control the measurand value.

Of necessity, the examples in Figure 1.1 do not include all types of processing, conditioning, and modes of use for instrumentation and control systems – that would be impossible in an introductory chapter of this nature.

Generally, the operating principles of input and output transducers are similar. However, their modes of operation are totally different – input transducers are typically used to convert measurand variations into weak electrical signals; output transducers typically convert strong signals into strong movements. For this reason we shall consider the two as totally separate types of device. This book is about input transducers: those transducers used for sensing purposes in electronic systems.

Terminology

To study transducers, some basic terms must be clarified first. Indeed, the very term *transducer* is not universally accepted to mean what we have already defined. A small number of engineers, for example, define a transducer as being only a device which converts an electrical signal into a physical quantity, i.e., what we have called an *output transducer*.

Sometimes the very fact that an input transducer must derive energy from somewhere to perform the conversion process (either from the measurand itself or from an external supply) helps us to define types of input transducer. For example, an input transducer which derives energy from the measurand is often called a **passive transducer**, although strictly speaking it should be known as a **sensor** (however, the term *sensor* has become almost synonymous with the more general term *transducer*). Following this convention, an **active transducer** is one which derives energy from an external power supply. Further, a **feedback transducer** is one used in a negative feedback control system to measure the signal which opposes the input signal in a balanced situation, rather than making a direct measurement of the measurand.

In general, of course, things are not as cut and dried as the previous paragraph would suggest and transducers tend to be known by the names which are common in a particular discipline. Often names are juxtaposed, too, between disciplines. Thus, names such as **transmitters, sensors, detectors, cells, gauges, pickups, probes**, as well as many suffixed derivatives ending in *-meter* such as **accelerometer, flowmeter, tachometer**, are all common and are all more-or-less correct. For convenience, we shall class them all as basic input transducers but call each *individual* transducer by the most common name it possesses, clarifying its function as required. Further, in general terms at least, a transducer will be taken to mean an *input* transducer while an output transducer will be called an output transducer. This is not so much an appreciation of the fact that there are many more types of input transducers than there are those of the output variety, merely that common usage of the term *transducer* has come to imply an input transducer. I feel no serious compunction to alter this implication.

The **accuracy** of a measurement, that is, the closeness of the measured value of a measurand to its actual value, is usually specified in the terms of **error**: the maximum possible difference between measured and actual values. For example, a 300 mm rule may have an error of, say, ± 1 mm. This means that the rule itself may have an actual length of somewhere between 299 mm and 301 mm – it *may* be exactly 300 mm, on the other hand it might not! Any measurement taken with the rule therefore has a maximum possible error of 1 mm, high or low. Sometimes, error is specified as a percentage. In the case of the rule, error could be specified as ± 0.0033 per cent. Sometimes, in special cases, the error is specified as a percentage of **full scale**, i.e., as a percentage of the maximum reading obtained. Errors can be a function of the transducer and equipment used, or can be user-generated. Accuracy can be affected by a number of factors, the important ones are discussed now.

Associated with accuracy, and often confused for it, is the **resolution** of a system, that is, the fineness with which the system can take measurements. Back to the rule example: if the rule is graduated in millimetres, then it should be possible to interpolate between 2 mm markings when measuring, to give a resolution of 0.5 mm. However, it is important to remember that although the resolution may be lower than the specified error, this does not mean that the reading has a lower error. The overall error is, in fact, greater.

The **sensitivity**, sometimes called the **scale factor** of a transducer is the change of its output as the input varies. For a **linear** transducer, say, an input transducer which produces an output voltage which changes linearly with changing temperature, the sensitivity may be quoted simply as the total output voltage range divided by the total input range – in the example given here it could be 10 V (the total output voltage range) divided by 100°C (the total input temperature range), or:

$$\frac{10}{100} = 0.1 \text{ V } °C^{-1}.$$

Linearity of a transducer may itself affect accuracy. It is often advisable to use a transducer with a linear response (Figure 1.2(a)), because the associated conditioning circuits are also linear and, hence, quite cheap to design and make. If a transducer is essentially nonlinear (Figure 1.2(b)), linearizing conditioning circuits may have to be used – these are costly and usually better to avoid. Fortunately by restricting use of many nonlinear transducers to only a small part of their total characteristic responses it is usually possible to ensure that fairly linear relationships between measurand and output signal are obtained.

On the other hand, a linear transducer is not advisable when the measurand changes in a nonlinear way. In such a case the linear transducer would only mirror the nonlinear change of measurand in its output signal.

Often, an essentially linear transducer may be operated out of its regular operating range and so it may **limit**, where the output signal abruptly levels off or saturates as the measurand exceeds full scale value. This in itself is a non-linearity.

In some cases a highly nonlinear response is preferred. In, say, an automobile exhaust gas monitoring transducer, a useful output signal would be one in which a particular signal is obtained when the exhaust emissions were acceptable, and another, distinctly different signal, is

Figure 1.2 *Linear and nonlinear transducers. (a) a linear transducer characteristic response; (b) a nonlinear transducer characteristic response. A nonlinear transducer can often be used over a restricted part of its total range to give an approximately linear characteristic response*

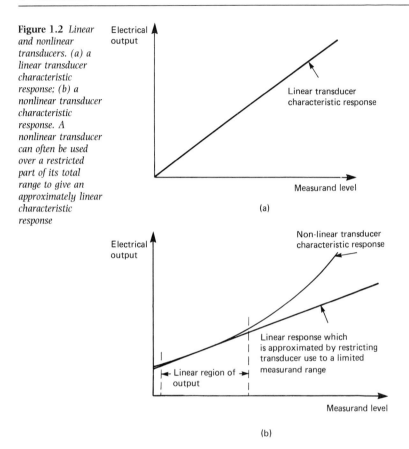

obtained if the exhaust emissions are unacceptable. Typically, one signal might be a logic 0 level and the other signal a logic 1 level. Obviously a 'snap' action switchover between the two signals is preferred.

The **hysteresis** of a transducer is an important factor. Taking again the exhaust gas monitoring transducer it may be that the exact point where the output signal changes from one state to another is different, depending on whether the noxious exhaust gas emission is increasing or decreasing. Figure 1.3 shows a possible characteristic of the transducer where the hysteresis effect is marked. As the noxious exhaust gas emission increases the transducer does not switch output states till gas percentage 2. When the level of noxious exhaust gas emission is decreasing the transducer does not switch states until gas percentage 1 is reached. Generally, this hysteresis effect should be as little as possible.

Repeatability of a transducer's output signal is a very important factor. The output signal should, ideally, be the same whenever the measurand value is the same. In some cases, particularly if the transducer exhibits a large hysteresis effect, the output signal may be different depending on which direction the chosen measurand value is approached.

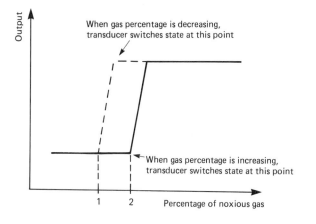

Figure 1.3
Illustrating hysteresis in a transducer

Another factor affecting a transducer's accuracy is its **response time**, which is the time taken for the output signal to respond to a change in measurand value. An instantaneous or step change in measurand value may not be accompanied by a simultaneous step change in output signal, if the transducer takes an amount of time to respond to the change. Such a transducer may produce no great error if the measurand changes only slowly or not at all, but the inherent inertia of the transducer's output means that the transducer cannot be used to follow rapid fluctuations in measurand value. This is not to say that every transducer used must have a response time faster than the changes in measurand value. In the case of, say, a transducer used to measure level of petrol in a car, a fast response time is a decided disadvantage – the driver does not wish to see a dashboard petrol gauge jumping about between full and empty if a bump on the road is driven over and the petrol is sloshing around the tank. There are other examples of transducers, too, whose response times must be carefully matched to applications – neither too fast, nor too slow – for accurate measurements to be possible.

The **bandwidth** of a transducer has a lot to do with response time. Measurand changes are accompanied by frequency components: according to Fourier analysis any waveform is made up of a combination of sinusoidal waveforms of various frequencies and relationships. The faster the measurand changes, the higher the frequencies are, and so the greater the **band** of frequencies present in the output signal. If the transducer bandwidth is relatively small, the higher frequency components present in the measurand value changes are not present in the output signal and a slow response time is the result.

Hostile environments

After all the preceding factors have been noted and the engineer begins to select a transducer for a particular application, it is then important to consider the application itself. Where is the transducer to be used? What conditions is it likely to encounter while it does its job?

This is a vitally important part of transducer choice, because the environment which the transducer is situated in can affect the transducer and its operation dramatically. **Hostile environments** must be foreseen by the engineer so that the transducer will function correctly, not just on commission, but for the measurement system's whole life.

Broadly speaking, there are three ways in which an environment can affect the transducer and its associated measurement system. First is the immediate effect of the environment on the transducer. Perhaps the temperature around a pressure transducer is high enough to melt parts of the transducer. Perhaps a temperature transducer measuring engine temperature cannot withstand the vibrations which occur in the engine's normal operation.

Second, given that the transducer is not immediately damaged by the hostile environment, is the transducer's accuracy maintained in the longer term? Or will the environment affect the operation of the transducer such that over a period of time its accuracy is degraded to a level which was unacceptable at the time of commission. In reality, as long as the transducer is renewed when there is a chance of degraded accuracy, this is a problem which may be sidestepped.

Finally, although this affects the transducer only indirectly, how does the environment affect the transmission path between the transducer and the rest of the measurement system? Can *it* withstand the hostile environment? And, is the environment likely to cause degradation of the signal?

Transducer make-up

Any transducer can be considered to be made up in the block diagram form of Figure 1.4. Here, a **sensing element** detects the quantity of measurand which is to be measured and converts it into another physical property. Then, a **transduction element** converts the physical property into an electrical signal so that the magnitude of the electrical signal is indicative of the measurand quantity. Other possible parts of transducers are integral signal conditioning and excitation circuits (see Chapter 3).

Often, the sensing element is a mechanical arrangement which simply converts a measurable part of the

Figure 1.4 *Block diagram of a transducer, showing parts common to all transducers. Those parts shown in broken line may not be required*

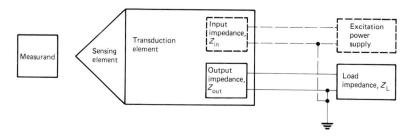

measurand quantity into a quantity which can be measured by the transduction element. Looked at in this way, the sensing element itself can be considered (in the strict dictionary sense only) a transducer. Figure 1.5 shows the principle of a device known as a **linear variable differential transformer** (LVDT) used together with a spring as a transducer to measure weight. The sensing element (i.e., the element which converts the measurand (weight) into another physical property (position)) is the spring, which is

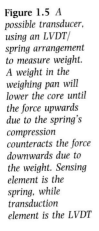

Figure 1.5 *A possible transducer, using an LVDT/ spring arrangement to measure weight. A weight in the weighing pan will lower the core until the force upwards due to the spring's compression counteracts the force downwards due to the weight. Sensing element is the spring, while transduction element is the LVDT*

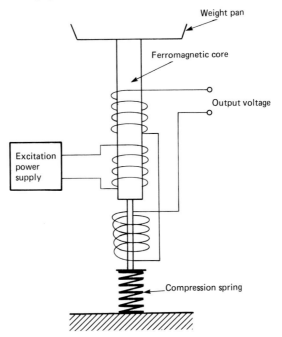

compressed when a weight is placed on the top of the transducer thus altering the position of the LVDT's central core. The transduction element is the LVDT itself, which converts the position of the core into an electrical signal whose magnitude is proportional to the weight being measured.

It is not always as easy as this example depicts, however, to identify separately the sensing and transduction elements of a transducer.

Analogue or digital?

There is a problem when incorporating transducers into many electronic systems. All transducers at their heart are of a purely analogue nature – largely due to the fact that there is no physical phenomenon which can be used directly to make a transducer with a digitally coded output signal. For an analogue system this situation presents few disadvantages. An increasingly large number of electronic systems are, however, digital. This means that, for an analogue transducer to be used in a digital system, the analogue output signal must be converted to a form which is acceptable to the digital system.

There are three main types of signals in electronic systems:

1 *Analogue*; in which the signal is an electronic representation or *analogy* (either voltage or current) of the original parameter.

2 *Digital*; in which a function, say, frequency, is used to represent the original parameter value.

3 *Coded digital*; in which a parallel digital signal, say, 8-bits wide, represents the original parameter value.

These three categories of signals usefully give categories of transducer, too. There are transducers whose outputs are purely an electronic analogue of the measurand value; there are transducers whose outputs are a digital representation of the measurand value; and there are transducers whose outputs are a coded digital representation. It is worth reiterating, however, that all types function by analogue principles. Also, there are only a few analogue principles which are used to make all types of transducer. Finally, electronic conversion of one type of signal to another is possible.

Book layout and description

There is a tremendous range of available transducers. There are, fortunately, only a few underlying physical principles by which these transducers function. By discussing these physical principles early on (Chapter 2) I hope to give the reader a general understanding of most categories of transducer before individual devices are considered. Further, discussions of transducer *operational* principles (Chapter 3) should give greater insight into their use. It is only then when individual devices are itemized (Chapters 4 to 8) in a reference-type format.

Main groups of measurands, and their associated quantities – which may themselves be measurands – are discussed individually, in the following order:

- Thermal measurands.
- Solid measurands.
- Fluid measurands.
- Acoustic and optical measurands.
- Chemical measurands.

One chapter is devoted to each main group of measurand. In each chapter the methods used and the basic transducers are described, with respect to the underlying transducer principles seen in Chapter 2. The reader is assumed to have a fundamental knowledge of the measurands and, so, little attempt at definitions of their principle is undertaken. New transducers and types of transducer are introduced regularly and if any transducers are missing, it is only because the information was not available at the time of compilation.

Finally, Chapter 9 discusses interfaces and connections to transducers, signal conditioning, analogue-to-digital conversion etc.

The book is not a guide on which transducer to use in a particular application. Often the engineer is faced with a choice of many transducers. The information I have presented here is a reference book of how transducers work and what main types of transducer are available, allowing the reader to:

1 Compare individual transducers (either input or output) and make an informed choice; and
2 Use the transducer and system to the best advantage.

2
Transducer principles

In each of the transducers considered in this book, the transduction element relies on a physical principle which affects the electrical characteristics, such that a change in measurand value alters the characteristics. It is this change in electrical characteristics which creates an electrical signal, dependent on the measurand.

Although there are many thousands of available transducers on the market, there is only a handful of physical principles. It is possible because of this to look at these principles in a reasonably straightforward manner. The main physical principles used can be grouped into only eight categories. These categories, in alphabetical order, with any subcategories, are detailed now.

Capacitive transducers

Capacitive transduction elements convert a change in measurand value into a change in capacitance. A capacitor is formed by two plates separated by a layer of dielectric, and its capacitance is given by the relationship:

$$C = \varepsilon \frac{A}{x}$$

Where ε is the permittivity of the dielectric, A is the surface area of each plate, and x is the distance between the plates.

From this relationship, a change in capacitance occurs with a variance in dielectric permittivity, plate surface area, or plate separation (Figure 2.1).

The capacitance of a capacitive transducer is usually measured:

1 Using an alternating current bridge circuit, where the transducer forms one arm of the bridge, or
2 Using the transducer as a frequency determining capacitor in an oscillator circuit.

Figure 2.1
*Capacitive
transduction. Two
plates separated by a
dielectric form a
capacitor, whose
capacitance depends
on the dimensions of
the plates and the
permittivity of the
dielectric*

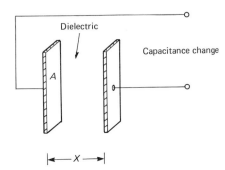

Piezoelectric transducers

One capacitive transduction principle which requires special note is the **piezoelectric effect** (Figure 2.2), in which a change in measurand value is converted to a change in the electrostatic charge or voltage generated across certain materials when mechanically stressed. The stress is typically developed by compression, tension, or bending forces; directly by the measurand sensing element, or indirectly by a mechanical connection to the measurand sensing element.

Figure 2.2
*Piezoelectric
transduction. A
force of capacitive
transduction where
the charge or
voltage across the
piezoelectric crystal
varies with the
stress on the crystal*

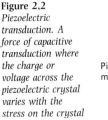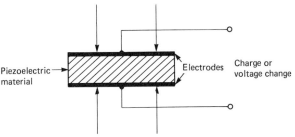

To utilize the change in charge or voltage, it is necessary to attach two metal electrodes to the piezoelectric material, effectively forming the plates of a capacitor, whose capacitance is given by the relationship:

$$C = \frac{Q}{V}$$

Where Q is the charge, and V is the voltage.

Piezoelectric material used in the construction of piezo-electric transducers is of three forms:

1 Naturally occurring crystals such as quartz and rochelle salt.
2 Synthetic crystals such as lithium sulphate.
3 Polarized ferroelectric ceramics such as barium titanate.

Electromagnetic transducers

An electromotive force (EMF) is generated across a conductor when a varying magnetic field cuts the conductor. Conversely, when a conductor moves through a

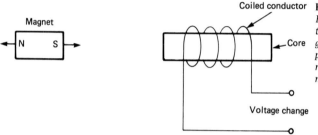

Figure 2.3
Electromagnetic transduction. EMF generated is proportional to the rate of change of magnetic flux

magnetic field an EMF is generated across the conductor (Figure 2.3). The EMF is given by the relationship:

$$E = -\frac{d(N\phi)}{dt}$$

Where $\dfrac{d(N\phi)}{dt}$ is the rate of change of the flux linkages.

Inductive transduction is shown in Figure 2.4, where the self-inductance of a coil is changed according to a change of measurand. The inductance changes can be made by the motion of a ferromagnetic core within a coil or by externally introduced flux changes in a coil having a fixed core.

Reluctive transduction is shown in Figure 2.5, where the reluctance path between two or more coils (or the

Figure 2.4
*Inductive
transduction.
Measurand value
alters the position of
the core within the
coil, so altering the
coil's self-inductance*

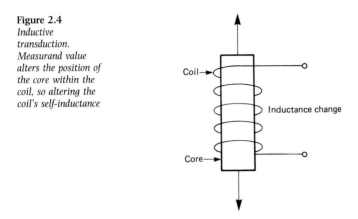

Figure 2.5
*Reluctive
transduction, used
in the LVDT to
change measurand
value into a voltage*

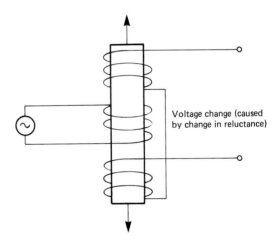

separated portions of one or more coils) is varied in accordance with a changing measurand value. As an alternating current is applied to the coil system, the change in measurand is transduced into a changing AC output voltage.

Electromechanical transducers

Electromechanical transducers are varied in make-up, but all have some form of mechanical contact arrangement, operated by measurand change. Usually, the contacts are of a simple make and break form such as in a bimetal switch (Figure 2.6). When the measurand value changes past the switchover point, the contacts either open or close, breaking or making an electrical circuit forming the output signal of the transducer.

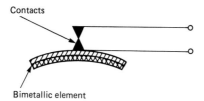

Contacts

Bimetallic element

Figure 2.6 *A bimetallic element, operating by electromechanical transduction*

Electromechanical transducers are essentially digital transducers because the make and break contacts form a two-state, on/off switching element.

Ionizing transducers

Ionizing transduction elements convert a change in measurand into a change in ionization current, such as through a liquid between two electrodes (Figure 2.7). A typical use of the ionization principle is in a device for measuring the acidity of a solution. A solution's degree of acidity is determined by the concentration of positively charged hydrogen ions – known as the *hydrogen potential*

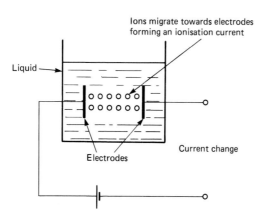

Ions migrate towards electrodes forming an ionisation current

Liquid

Electrodes

Current change

Figure 2.7 *Ionizing transduction, in which ions in the liquid migrate to each electrode, so acting as charge carriers and causing current flow* (more commonly abbreviated to pH value). The pH value is given by the formula:

$$pH = - \log [H^+]$$

Where $[H^+]$ is the hydrogen ion concentration in grams per litre.

Values of pH range from 0 for a strong acid solution, through 7 for a neutral solution (say, pure water), to 14 for a strong alkaline solution. Typical pH probes have an electrode held in a gel of known pH value, and an electrode formed by a special glass membrane which comes into contact with the solution whose pH value is to be measured. The potential difference between the two electrodes is indicative of the pH value of the solution (about 59 mV per unit of pH).

Photoelectrical transducers

Photoelectrical transducers are those which respond to the amount of electromagnetic radiation incident on the transduction element surface. The radiation may be of a visible nature, i.e., light, but often is of a longer or shorter wavelength and is invisible. There are three main types of photoelectrical transducers: two of which are officially classed as semiconductor devices (photovoltaic and photosemiconductors), so will be discussed in detail in the relevant section – although one particular type of photovoltaic transducer is not a semiconductor device and so will be discussed here.

Photoconductive transducers

These transducers convert a change in measurand value into a change of the resistance of the material used (Figure 2.8). Although the material used is a semiconductor, photoconductive transducers are not generally thought of as semiconductor devices, because they have no junction between different types of semiconductors. Also they are passive, i.e., requiring no external excitation. Indeed, their common names, for example **light dependent resistors**, indicate how the transducers are typically used.

Resistance of the material is a function of the majority charge carrier density, and as the density increases with the level of radiation, so the conductivity increases. As conductivity is the inverse of resistance, it can be seen that resistance is an inverse function of the amount of incident

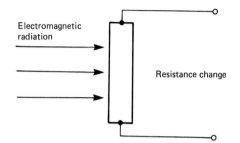

Figure 2.8 *Photo-conductive transduction. Electromagnetic radiation striking the photoconductive element causes a change in resistance*

radiation. Values of resistance at full illumination are commonly around 100 to 200 Ω, with a resistance of megohms in the dark. Common substances used in the construction of light dependent resistors are cadmium sulphide and cadmium selenide.

Solar cells

Solar cells are **photovoltaic transducers** which convert electromagnetic radiated energy to electrical energy, so that a change in measurand radiation value is converted to a change in output voltage (Figure 2.9).

Construction comprises a high resistance photosensitive layer of material, sandwiched between two conduct-

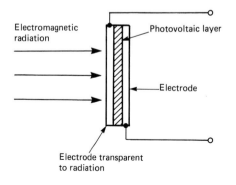

Figure 2.9 *A solar cell. One example of photovoltaic transduction*

ing electrodes. One of these electrodes is of a transparent material, so that incident radiation passes through it to the photosensitive material. Under full illumination the cell develops an output voltage of about 0.5 V across the electrodes.

Resistive transducers

By far the largest category of transducers is based on resistive principles, where a change in measurand value is converted into a change in resistance. Resistance changes can be caused by a number of effects in the transduction element, such as heating or cooling, mechanical stresses, light (as in the photoconductive transducer), wetting or drying, mechanical movement of a rheostat wiper.

Figure 2.10
Resistive transduction, used in a potentiometer arrangement to cause an output voltage change

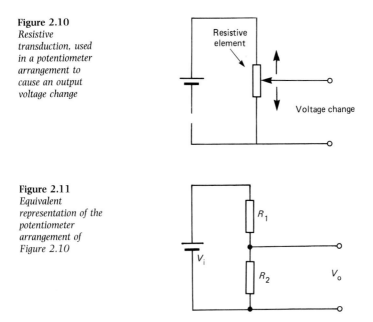

Figure 2.11
Equivalent representation of the potentiometer arrangement of Figure 2.10

If a fixed current flows through the resistive material at the time of the measurand change, the result will be a voltage change across the material which indicates the measurand change.

A version of resistive transduction is **potentiometric transduction** in which the measurand value change is converted into a voltage ratio change, by a change in the position of a wiper on a resistance element across which excitation exists (Figure 2.10). Some mechanical linkage converts the measurand value change into wiper displacement.

The simple potentiometer arrangement of Figure 2.10 may be represented by the circuit of Figure 2.11, where the output voltage is given by the relationship:

$$V_0 = \frac{V_1 R_2}{R_1 + R_2}$$

Where V_1 is the applied voltage input.

If the applied voltage is constant and the measurand value determines the position of the wiper, then the output voltage is a direct function of the measurand value.

Transducers can either use the potentiometric arrangement (having one or more resistances in a circuit which follows the arrangement) or they may be actual potentiometers. Make-up of the potentiometer transducer element is varied. Some transducers have a wirewound resistance, some have a cermet (metallised ceramic) substrate, some have a conductive plastic film resistance. Mechanically, potentiometers are obtained in which the full range of wiper positions is effected by a rotational movement of as little as 270° or so, while others which require 10 or even 20 complete turns (3600° or 7200°) or more are available.

Wheatstone bridge
By combining two potentiometer arrangements in parallel (Figure 2.12) the **Wheatstone bridge** is formed, which may be used for accurate measurements of resistance. The output voltage of the Wheatstone bridge is given by:

$$V_0 = V_1 \left(\frac{R_3}{R_3 + R_4} - \frac{R_1}{R_1 + R_2} \right)$$

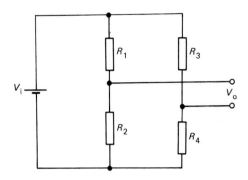

When correctly set up, the output voltage of the Wheatstone bridge should be zero, so:

$$\frac{R_3/R_4}{1 + R_3/R_4} = \frac{R_1/R_2}{1 + R_1/R_2}$$

Giving:

$$\frac{R_1}{R_2} = \frac{R_3}{R_4}$$

Strain gauges

Because the resistance of a conductor is given by:

$$R = \rho L/A$$

Where ρ is the resistivity of the material, L is the length, and A is the cross-sectional area, the resistance can be changed by any change of measurand value which affects one or more of the three dependants.

This is used to advantage in the **strain gauge**: a transducer which relies on an applied strain to change its resistance (Figure 2.13). The transducer is normally used within a Wheatstone bridge arrangement with either one, two, or all four of the bridge arms being individual strain gauges, so that the output voltage change is an indication of measurand (the strain) change.

Strain gauges are available which use metal transduction elements, and application of strain simply changes their length and cross-sectional area to alter the resistance value. Certain substances, however, notably semiconductors, exhibit the **piezoresistive** effect, in which application of strain greatly affects their *resistivity*. Strain gauges of this type have a sensitivity of approximately two orders greater than the former type.

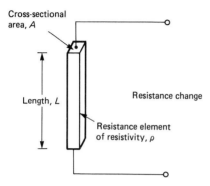

Figure 2.13 *Strain gauge transduction. Applied stress or strain causes the element to vary in length, so causing a change in resistance*

Generally, any parameter which produces a motion or a force can be used to develop a strain gauge transducer.

Resistance can also vary with *temperature* and, for a metal, may be specified according to the simplified linear relationship:

$$R = R_0(1 + \alpha T)$$

Where R_0 is the resistance at 0°C, T is the temperature in degrees Celsius, and α is the *temperature coefficient of resistance*.

Typical resistance against temperature characteristics for a number of metals are shown in Figure 2.14, and a high degree of linearity is apparent. Usually, platinum wire is the material used to make temperature transducers of this type.

Thermistors

Another main group of temperature sensing transducers is formed by those devices known as **thermistors**. They have highly nonlinear characteristics, but can be used very effectively in temperature measurement systems. The

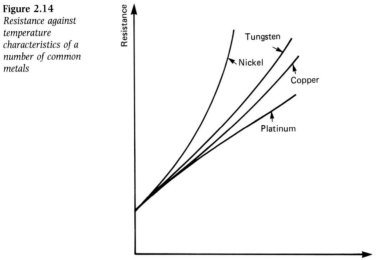

Figure 2.14
Resistance against temperature characteristics of a number of common metals

resistance of a thermistor may be expressed by the relationship:

$$R_{\mathrm{T}} = A \exp \frac{B}{T}$$

Where R_T is the resistance at T degrees Kelvin, and A and B are constants for the material (B is the **characteristic temperature** of the device).

A typical characteristic is shown in Figure 2.15, where the exponential nonlinear curve of the thermistor can be seen. Compared with the characteristics of resistance transducers (Figure 2.14) that of the thermistor:

- Is much steeper, i.e., the temperature coefficient of resistance is much larger than that of the metals, at least over the important part of the curve.
- Decreases rather than increases with increasing temperature, i.e., the temperature coefficient of resistance is *negative*.

These thermistor resistive transducers with negative temperature coefficients are commonly known as NTC thermistors. At this point it is worth mentioning that there are thermistors available which have a *positive* temperature coefficient of resistance (known as PTC thermistors), but these are used mainly in applications such as overheat protection, rather than for temperature measurement.

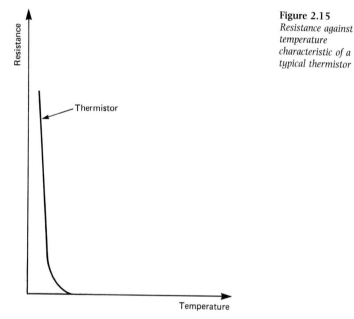

Figure 2.15
Resistance against temperature characteristic of a typical thermistor

An alternative and more convenient expression for the thermistor formula – if the resistance is known at some temperature T_1 – where the resistance at T_1 is given by:

$$R_1 = A \exp \frac{B}{T_1}$$

Can be found by dividing the previous expression by this one, to give:

$$R = R_1 \exp B\left(\frac{1}{T} - \frac{1}{T_1}\right)$$

Thermistors are very much smaller than metallic resistance transducers, and so can respond more rapidly to temperature changes. Their size, on the other hand, means

that low currents must be used in order that self-heating
due to the current does not affect the measurement.

Semiconductor transducers

Semiconductor devices are electronic components made
from the category of elements known as semiconductors.
Pure or *intrinsic* semiconductors are not generally used in
the devices, but are first *doped* with impurities into the
semiconductor crystal lattice, to become *extrinsic*
semiconductors.

Extrinsic semiconductors can be doped in such a way as
to have an excess of electrons (n-type semiconductor) or a
reduction in the number of electrons (p-type
semiconductor). It is the presence of impurities within the
semiconductor crystal lattice which defines the extent to
which the lattice conducts electricity.

A single layer of n- or p-type semiconductor is of no
purpose, and it is only when two or more layers of the
different types are joined that the semiconductor material
becomes a semiconductor device. The simplest p-n junction forms a rectifying device or diode, and the diode
current against voltage characteristics may be stated by
the relationship known as the *Schockley equation*, also
known as the *ideal diode equation*:

$$I = I_o \left[\exp \left(\frac{qV}{kT} \right) - 1 \right]$$

Where I_o is the saturation current (also known as the
leakage current), q is the electron charge, V is the applied
voltage, k is Boltzmann's constant, and T is the temperature in degrees Kelvin.

Any measurand change which changes any of the
conditions of the above relationship can, of course, be used
to change the current flowing through the junction. For
example, diodes may be used as a temperature transducer
because the saturation current of the semiconductor
varies with temperature. The saturation current of silicon

is about 25 nA at 25°C, but increases to about 6.5 mA at 150°C.

Photodetectors

Semiconductor transducers which can be used to measure a change in measurand value of light are known generically as **photodetectors**. The **photovoltaic transducer** is the simplest of these and is a type of semiconductor diode. Several types are available. The most basic is known as the **photodiode**, which utilizes the effects of incident light (visible or of other wavelengths) on a reverse-biased p-n junction to alter the amount of current flowing through the junction. Such a photodiode will have a dynamic response of only a few nanoseconds.

For an even faster response, however, a common photodiode called a PIN diode is manufactured with a layer of intrinsic semiconductor between the p- and n-type layers. This improves sensitivity to incident light, and also lowers the capacitance at the junction so that the photodiode can respond to more rapid changes in measurand level.

Phototransistors

In many applications, photodiodes are used in conjunction with an amplifier to increase sensitivity. However, as a conventional transistor (a three-layer semiconductor device – n-p-n or p-n-p) contains a reverse-biased p-n junction and has the ability to amplify current, it possesses everything necessary for a photodiode and an amplifier – all in one package. **Phototransistors**, as opposed to conventional transistors however, are manufactured in a transparent package to allow passage of light. Light falling on a phototransistor's collector-base junction (a reverse-biased p-n junction) causes a base photocurrent to flow, which is amplified by the transistor's gain to produce a, potentially, much larger emitter current.

The emitter current of the phototransistor is given by the relationship:

$$I_E = (1 + h_{FE}) I_P$$

Where h_{FE} is the transistor's DC current gain, and I_P is the base photocurrent.

For even greater gain, **photodarlington detectors** are available, which comprise a phototransistor and a high gain transistor in Darlington pair mode, housed in a single package. As photodetectors *are* semiconductor devices, they have a temperature dependent saturation current, so even with zero incident light a **dark current** flows which sets a limit to the device's abilities to measure low light levels.

Hall effect

When a current carrying conductor is placed in a magnetic field such that the current and field are perpendicular, a transverse electric field is developed which is proportional to the product of the field's magnetic flux density and the current. This effect is present in all conductors but is only of significance in semiconductors, and is known as the **Hall effect**.

Figure 2.16
Illustrating the Hall effect, in which an electric field is set up perpendicularly to an applied magnetic field and current. Present in all conductors but only of significance in certain semiconductors

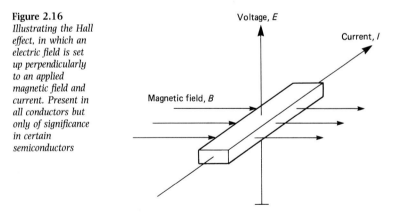

Figure 2.16 shows a wafer of semiconductor material, with an applied magnetic field, B, perpendicular to a current, I, and the resultant electric field, E. The relation-

ship between magnetic field, current and electric field is given by:

$$E = -R_H(I \times B)$$

Where R_H is the **Hall coefficient** and is equal to:

$$\frac{1}{ne}$$

Where n is the number of charge carriers per unit volume which make up the current, and e is the charge on the charge carriers.

The Hall effect is used in many transducers, particularly relating to measurements of magnetic field, but can also be used as the basis of a noncontact switching device.

Thermoelectric transducers

Thermoelectric transduction elements convert a change in measurand (temperature) into a change in the current generated by a temperature difference between the junctions of two dissimilar materials, using the *Seebeck effect* (Figure 2.17).

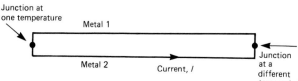

Junction at one temperature

Metal 1

Metal 2   Current, *I*

Junction at a different temperature

The thermoelectric transducer is more commonly known as a **thermocouple**, in which a probe comprising one junction is located at the point whose temperature is to be measured, while the other junction is located at a point of reference temperature (Figure 2.18). The potential differences which occur at the two junctions (known as **contact potentials**), V_1 and V_2, depend on the junction temperatures. The voltmeter measures the difference between these two contact potentials:

$$V_1 - V_2$$

So the voltmeter reading is an indication of the difference in temperature between the junctions.

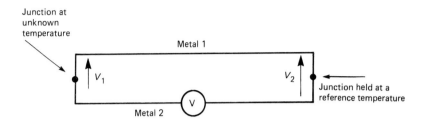

Junction at
unknown
temperature

Metal 1

V_1

V_2

Junction held at a
reference temperature

V

Metal 2

Figure 2.18
*Principle of the
thermocouple. One
junction is
maintained at a
reference
temperature, while
the other junction is
placed at the
temperature to be
measured. The
voltmeter displays a
reading which is
indicative of the
temperature
difference*

Figure 2.19 shows a number of curves of voltage against temperature difference for several typical thermocouple junction materials. Although seemingly linear, checking with a ruler shows they are not quite.

In practice, accurate temperature measurement using a thermocouple may not be as simple as the above discussion might suggest, because connection of the voltmeter to the thermoelectric circuit itself forms new junctions in the circuit, the temperature to be measured may be a distance from the voltmeter and it may be difficult to create a sufficiently stable reference temperature.

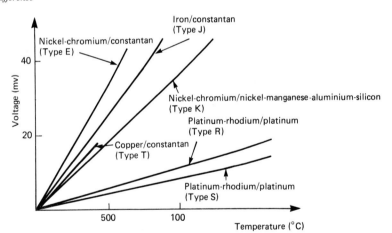

Figure 2.19
*Typical curves of
voltage against
temperature
differences for a
number of common
thermocouple
junction materials*

3
Transducer characteristics and engineering considerations

Generally, a transducer is designed to sense one specific measurand, and to respond only to that measurand. So, a thermistor, for example, is designed to measure temperature change. Often, however, other measurands can be calculated by their known relationship with the measurand sensed by the transducer: velocity, say, can be calculated from a measurement of displacement, simply by dividing the displacement by the time over which the displacement takes place. The functions for which a transducer can be used, therefore, sometimes depend on external factors of system design, and not simply the internal transduction factors. Internal transduction factors are still of great relevance, however, and must be considered in any serious study of transducers.

Transduction factors
The **range** of a transducer is specified by the upper and lower limits of the measurand values. A **unidirectional** range is one in which either positive or negative measurand values may be encountered (say, 0 to 10 kg). A unidirectional range may be **expanded** such that no zero measurand value is measured (say, 120 to 900 rpm). A **bidirectional** range is one where positive and negative values are encountered (say, $\pm 60\,^{\circ}$C, in this example symmetrically, or asymmetrically, say, -20 to $+100\,^{\circ}$C).

In black box terms (Figure 3.1, a repeat of Figure 1.4) the transducer is a fairly simple component. It has a

Figure 3.1 *A repeat of Figure 1.4, illustrating in block diagram form the make-up of a transducer*

sensing element which responds directly to the measurand, a **transduction element** (operating on one of the basic transduction principles discussed in Chapter 2) which generates some electrical output in response to the value of measurand, and possibly some internal excitation and/or signal conditioning circuits.

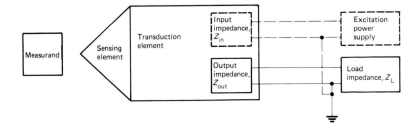

Two connections to the transducer are an excitation supply (which may be internal, or, may not exist at all) and an output to a load. **Excitation** is required by a transducer (with the exception of passive transducers) for its correct operation, and may be a voltage or current. The impedance of the excitation supply, Z_s, is known as the **source impedance**; the impedance of the transducer, Z_{in}, presented to the excitation supply is the **input impedance**. Any impedance due to the presence of cable between the excitation supply and the transducer is always considered to be part of the source impedance. The **output impedance**, Z_{out}, is the impedance across the output terminals of the transducer; while the impedance presented to the output terminals of the transducer is the **load impedance**, Z_L. Similar to the excitation cable, any impedance of cable between transducer and load is always considered to be part of the load impedance. Matching of a transducer to a measurement system requires careful consideration of these impedances.

Excitation supply and output signal may be totally isolated in electrical terms, or may have a common return. The return connections are usually electrically isolated from the transducer's case and can be earthed or floating, depending on the earthing arrangements used in the measurement system. Mechanically, transducers initially depend on the transduction element used. However, as the transduction element must usually be housed in a body of some sort, the body itself must be considered. The mechanical design of a transducer body allows three main functions:

1 Installation and handling of the device.
2 Prevention of damage by the measurand or the environment.
3 To allow correct interface with the measuring system.

As well as choosing a transducer which performs to a sufficient electrical standard, the engineer must decide whether a selected transducer performs these mechanical functions, too. It is pointless, say, to measure the temperature of nitric acid by inserting a metal-bodied temperature transducer into the fluid. More about this aspect, later.

Transducer performance is an important factor. Generally, there are four main areas of a transducer's performance which the engineer needs to consider:

1 **Static operation**: performance of the transducer at room conditions, with little or no changes in measurand value, and without mechanical movement (unless movement is the measurand, of course). Room conditions are difficult to define, but may be taken to be: a temperature of $25\,^{\circ}$C; humidity of 90% or less; pressure of 100 kPa $\pm$ 10 kPa. A number of the transducer's characteristics are important here, and have been formally defined in Chapter 1: accuracy; error; linearity; sensitivity; hysteresis; repeatability.
2 **Dynamic operation**: performance of the transducer at room conditions, with fast changes of measurand value, but without mechanical movement. One of the important characteristics which concerns dynamic operation, already defined in Chapter 1 is response time, i.e., the time taken for the output signal of the transducer to

respond to a change in measurand value. It is important to realize that whatever transducer is used to measure a measurand, the output signal at any instant of time will represent the value of the measurand *at some earlier instant of time*. In other words, the output signal will *always* suffer from a delay, however small, between the measurand change occurring and the correct representation. The delay is defined generally by the transducer's response time and associated terms.

When considering transducer response times, it is convenient to think of the measurand change occurring as a **step**, i.e., changing from one value to another, instantly.

First-order linear response

Figure 3.2 *Step response of a first-order linear transducer. After a period equal to five time constants the output may be assumed to be at its final value*

Figure 3.2 shows a graph of a step change in measurand value and the resultant output signal of a possible transducer. A number of things are of note here. First, the curve (called a **step response** or **transient response**) is an exponential one, so the time the output takes to rise from the initial level to 63.2% of the final value is known as the **time constant** and is assigned the special symbol, τ. Next,

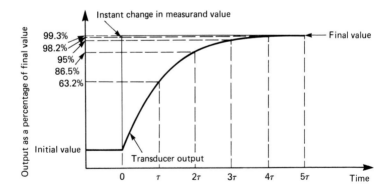

after a time equal to two time constants the output has gone to 86.5% of the final value; after three time constants the output is 95% of final value; after four time constants it is 98.2% of final value; and after five time constants,

Table 3.1 *Step response of a first-order linear transducer, tabulated as a percentage of final output value against time constants*

Time interval	Output value as a percentage of final value
1	63.2
2	86.5
3	95.0
4	98.2
5	99.3

99.3%. In fact, what is happening is that the output value, after each time constant, approaches the final value by 63.2% of the *difference* between the value at the previous time constant and the final value. The output values as a percentage of final value at each time constant are tabulated in Table 3.1

Because the curve *is* exponential which means that the output value, in theory at least, never actually reaches the final value, and because the value after the five time constants is within 1% of the maximum possible value, it is usually considered that after 5 τ the output value is *equal* to the final value. However, this is an engineering rule-of-thumb, no more, no less. The time taken for the output to rise from 10% to 90% of the final value is known as the **rise time**.

An important point about transducers which follow this **first-order linear response** is that the response time can be totally characterized by the time constant, τ.

Second-order linear response
Figure 3.3 shows the step response of another possible transducer for a step change in measurand value. This is a totally different response from that of the previous trans-

Figure 3.3 *Step response of a second-order linear transducer*

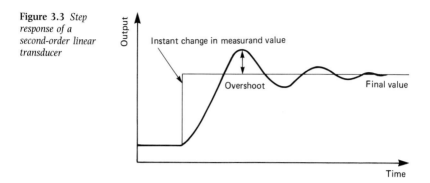

ducer, and shows that the output value oscillates around the final value, at a frequency known as the **ringing frequency**, before coming to rest at that value. The maximum value by which the output signal exceeds the final value is the **overshoot** and the maximum value of overshoot occurs at the peak of the first oscillation. If the transducer's sensing element is set into free oscillation, the frequency of this is known as the **natural angular frequency**, ω_n, not necessarily the same as the ringing frequency.

Figure 3.4 *Illustrating the effects of damping on the step responses of second-order linear transducer outputs*

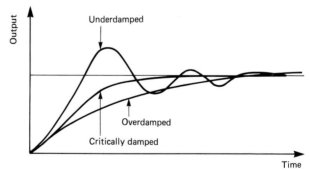

The time for which the output value oscillates around the final value, and indeed whether it oscillates at all, is determined by the amount of **damping** which is applied to the transducer. Figure 3.4 shows the step responses for a transducer for three types of damping: **underdamping**

(which gives the step response similar to that in Figure 3.3); **overdamping** (which gives a step response which does not oscillate but takes a considerable time to reach the final value); **critical damping** (which gives a step response which does not oscillate and reaches the final value in the shortest time). Needless to say, critical damping is usually taken to be the ideal for such a transducer.

The ratio of the actual damping to the degree of damping required for critical damping is the **damping ratio** or **damping factor** and is given the symbol b in most texts. So, a damping ratio of 1 indicates critical damping, damping ratios over 1 indicate overdamping, and damping ratios less than 1 indicate underdamping.

Transducers of this type are said to have **second-order linear responses**. Whereas first-order linear transducer response times can be characterized by the single quantity of time constant, second-order linear transducer response times require two quantities; damping ratio b, and natural angular frequency, ω_n, to give full characterization. It is a little more complicated to calculate the second-order linear response because of this but, fortunately, it can be done without maths if *normalized* curves of response are drawn, i.e., with normalized axes. Figure 3.5 shows such a set of curves where the response value is calibrated as the step response, x_r, expressed as a fraction of the final value, x_f. Similarly, time, t, is expressed as a fraction of $1/\omega_n$.

Normalized second-order linear response curves apply to *any transducer with a second-order linear response*. So, from the damping ratio, the engineer simply selects the corresponding curve then sets the timescale from the natural angular frequency.

First- and second-order linear transducers
Both first- and second-order linear transducers are common and their main differences, shown in the previous discussion, should be known to the engineer. Which type is used in an application depends largely on the measurand.

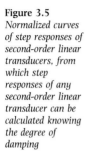

Figure 3.5
Normalized curves of step responses of second-order linear transducers, from which step responses of any second-order linear transducer can be calculated knowing the degree of damping

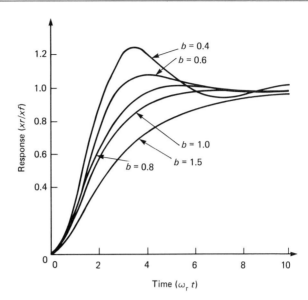

What defines a first-order linear transducer as different from a second-order linear transducer is merely that the mathematical representation of the transducer's behaviour is a first-order differential equation, i.e., the highest differential with respect to time in the equation is itself first-order. Thus, if the differential equation which describes the behaviour of a transducer is of the form:

$$a = K\frac{dx}{dt}$$

then the response is of first-order, because the highest differential with respect to time (indeed, the only differential with respect to time) is dx/dt, a first-order term.

On the other hand, if the differential equation describing the transducer behaviour is of the form:

$$a = \frac{d^2x}{dt^2} + 2b\omega_n\frac{dx}{dt} + \omega_n^2 x$$

then the response is of second-order, because the highest differential with respect to time is d^2x/dt^2 – a second-order term.

3 **Environmental aspects:** when a transducer is operated with a varying measurand, and when mechanical movements or other effects due to the environmental conditions apply. Most transducers are used in far from ideal environmental conditions. Room temperature, humidity and pressure are rarely experienced in industry and so a chosen transducer must be known to be able to function correctly, or at least within specified tolerances, under all environmental conditions which it is likely to experience.

Effects of temperature must be known and are generally specified by a manufacturer for any transducer, in order that the engineer can design any necessary circuit compensation, or correct final data from the measurement system. Similarly, vibration, acceleration, ambient pressure variations, or incorrect mounting of the transducer can all affect measurements.

A number of other environmental effects depend on how the transducer is used. Immersion in liquid may affect operation if the transducer is not sealed for the purpose, corrosion by salt, acid, etc., may occur if the body of the transducer cannot withstand such environments. Local electromagnetic fields may affect the transducer and its interfacing circuits.

4 **Reliability:** the ability of a transducer to operate correctly, under known static, dynamic and environmental conditions, for a certain period of time. How long the transducer operates correctly for is more precisely defined in terms of **failure**, i.e. how long the component functions before it fails.

The mechanism or process which causes the failure may be of a number of types, and to varying degrees. Some common types are:

● **Sudden failure**; where the component fails without warning and so cannot be predictable.
● **Gradual failure**; where failure may be possible to predict

if examination shows a discrepancy between specified and actual results.

- **Partial failure**; where the transducer is still capable of operating, even if the results are out of specification.
- **Complete failure**; where no operation is possible.
- **Catastrophic failure**; where the failure is both sudden and complete.
- **Degradation failure**; where the failure is both gradual and partial.

Failure of a component may be because of misuse, i.e., it has been subjected to conditions beyond its capabilities, or may be due to an inherent weakness, causing it to fail within its stated capabilities.

With time, all components fail. Obviously, it is impossible to define precisely when any particular component will fail so, instead, manufacturers will give an indication of the *likelihood* of a failure occurring. There are a number of ways this can be done.

Exactly *when* a particular component is likely to fail may be stated as the **mean time between failure** (MTBF), where the MTBF is stated for a family of identical components as:

$$\text{MTBF} = \frac{\text{the total operating time (in hours)}}{\text{the number of times operation is interrupted}}$$

Where a component cannot be repaired, the likelihood of failure may be stated as the **mean time to failure** (MTTF), given by:

$$\text{MTTF} = \frac{\begin{array}{c}\text{lifespan of each}\\\text{failure}\end{array} + \begin{array}{c}\text{total period}\\\text{of use}\end{array} + \begin{array}{c}\text{number of}\\\text{survivors}\end{array}}{\text{the number of times operation is interrupted}}$$

The lifetime of a family of components is sometimes graphically represented in a **bathtub curve**, called because of its shape (Figure 3.6). Three distinct areas are clearly identifiable on this graph.

When a component is first operated, failure rate is quite high; failures during this period (often called the **burn-in** period) are called **early failures**. Later on, during what is called the **useful life period**, failures are less common, with a fairly constant failure rate. During the **wear-out period** the component is reaching the end of its life and so failures occur at a higher rate again, and are called **wear-out failures**.

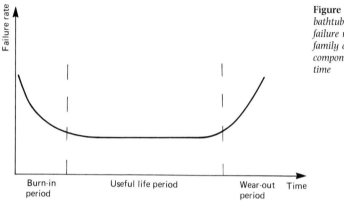

Figure 3.6 *A bathtub curve of failure rates of a family of components against time*

Economic factors

Use of high technology silicon products has drastically reduced the cost of all electronic systems, giving the added benefits of greater reliability, ease of maintenance, and general size reduction. Alongside this, transducer technology is steadily improving, too, and modern miniature transducers complete with integral signal conditioning circuits are frequently appearing, with improved performances over previously available counterparts. In terms of cash layout, these devices may not be cheaper than the older ones, but taking into account the use of internal signal conditioning circuits, system cost still may be lower than previously possible.

In the future, integration of signal conditioning circuits will continue to the extent where transducer output is a direct digital, fully-coded signal, suitable for direct interface to microcomputers and the like. Presently, a few such transducers exist, and there are more to come. Whatever happens in this respect, there will always be a justification for analogue transducers – digital systems will never be cheap enough to eliminate the need for certain analogue systems and some applications, it is argued, can never be replaced by digital techniques, anyway.

To the engineer, system cost is partly within engineering control: a number of solutions will exist for any viable product. In order to be fair, the engineer must look at all possible alternatives: digital; analogue; part-digital/part-analogue, and design the system with an overall idea of the costs involved. Often the transducer is a key component in the system. If, say, the system can only function with a highly accurate, highly reliable and highly sensitive transducer (which will be, no doubt, highly expensive, too), then the cost of the remaining parts of the system could well be little in comparison with that of transducer. On the other hand, if the transducer requirements are simple and few then perhaps any old transducer will do – system cost in such an example depends largely on the remaining parts of the system, not on the transducer.

Economic considerations like these, of course, apply not only to the selection of transducers, but to any product. Further, they are far too complex and there are far too many of them to be covered completely in a book of this nature.

4
Thermal measurands

The basic quantity covered in this chapter is temperature. A body's temperature, however, can be used to indicate that body's presence, so one method of proximity sensing is covered here, too.

A body or system's **temperature** is its thermal state: it is a measure of the kinetic energy due to heat agitation of the body or system's molecules – the potential of heat flow. **Heat** is energy due to temperature difference between a body or system and its surroundings.

Heat may be transferred from one body or system to another by one or more of three **heat transfer** methods:

1 **Conduction:** diffusion through a medium.
2 **Convection:** movement of a medium.
3 **Radiation:** by electromagnetic waves.

Heat capacity is the quantity of heat required to raise the temperature of a body or system by one degree of temperature. **Specific heat** is the ratio of the heat capacity of a body to the body's mass.

Thermal resistance is a measure of a body's ability to prevent heat flow through it. **Thermal equilibrium** is the condition between a body or system and its surroundings, when no heat transfer occurs between them.

Boiling point of a substance is the temperature of equilibrium between its liquid and vapour states. **Freezing point** is the temperature of equilibrium between the solid and liquid phases. In terms of water, the boiling point is 100 ° C and the freezing point, or **ice point** is 0 ° C. All these points are applicable only at the *standard atmospheric pressure* of 101 325 Pa.

Table 4.1 *Eleven main points of the International Practical Temperature Scale (IPTS) and the standard instruments and temperatures*

Point	Temperature	Instrument
Freezing point of gold	1337.58 K	Optical pyrometer (above 1337.58 K)
Freezing point of silver	1235.08 K	Thermocouple (903.87 K to 1337.58 K)
Freezing point of zinc	692.73 K	
Boiling point of water	373.15 K	
Triple point of water	273.16 K	
Boiling point of oxygen	90.188 K	
Triple point of oxygen	54.361 K	
Boiling point of neon	27.102 K	Platinum resistance thermometer (13.81 K to 903.89 K)
Boiling point of equilibrium hydrogen		
Equilibrium between the liquid and vapour phases of equilibrium hydrogen at 33330.6 Pa pressure	20.28 K 17.042 K	
Triple point of equilibrium hydrogen	13.81 K	

Temperature scales

The Celsius temperature scale, i.e., measured in °C, is only one of many used to measure temperature. It was originally defined by the freezing and boiling points of water. The **thermodynamic scale** defined in the SI (Système International) system of units, has different definition points: **absolute zero** (0 K – note that temperatures on the thermodynamic scale are quoted in degrees kelvin, but have no degree symbol) the theoretical minimum temperature of any substance; and the **triple point of water** (273.16 K). The triple point of water is the temperature of equilibrium at which water, ice and steam can coexist. On the thermodynamic scale, 0 °C at standard atmospheric pressure is 273.15 K.

The use of 100 discrete and equal steps in temperature between the boiling and freezing points of water at standard atmospheric pressure, to define the Celsius temperature scale, is purely arbitrary. It is just as arbitrary, of course, to use 180 discrete steps as in the Fahrenheit temperature scale. However, as 100 steps at least give the feeling of decimalization, the Celsius temperature scale is recognized internationally, along with the thermodynamic temperature scale, in the International Practical Temperature Scale (IPTS), recommended by the International Committee on Weights and Measures, in 1968. The IPTS is based on the assigned values of the temperatures of a number of equilibrium states, and on standard instruments calibrated at those temperatures.

Use of temperatures *and* standard instruments in the IPTS ensures that any discrepancies in the measurement of temperature, due to different types of instruments and measuring devices, can be minimized. For spaces in between the reference points and instruments, interpolation equations ensure that accuracy is maintained. Table 4.1 lists the eleven main fixed points of the IPTS and the standard instruments to be used, along with assigned temperatures.

Platinum resistance thermometers
The resistance of a wire or film of platinum is used as an indication of temperature in these transducers, often known as **resistance temperature detectors** (RTD). This is not to say that other metals cannot be used, in fact some are, but the vast majority of these transducers use platinum as the sensing element.

Sensitivity of resistance temperature transducers is quite low and dynamic response (because of the construction) is generally slow. They can be damaged easily by vibration or shock.

Resistance versus temperature characteristic is fully defined in BS 1904: Industrial platinum resistance thermometer elements for a temperature range of $-220\,°C$ to $+1050\,°C$.

There are two major types of platinum wire resistance transducers: immersion probes and surface-mounted sensors. Wire elements are typically wound on a ceramic

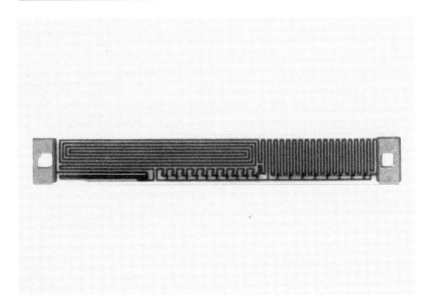

former, with the minimum of tension, and generally sheathed in a casing material to protect it from the measuring environment. Construction of a typical platinum wire element probe is shown in Figure 4.1(a) and the construction of a surface mounted platinum temperature sensor is shown in Figure 4.1(b). Film element transducers which use a metal film deposited onto an insulating substrate, are not so common as wire types, although thin-film types are becoming increasingly popular, due to their small size, improved dynamic response and sensitivity, and relative cheapness. Construction of such a transducer is shown in Figure 4.1(c).

In use, the platinum resistance transducer would normally be connected to form one arm of a Wheatstone bridge measuring arrangement, which allows high measurement accuracy. However, the transducer's quite low resistance (about $100 \ \Omega$) creates an instrumentation problem, as the resistance of the connecting leads between transducer and the measuring circuit may be significant (see Chapter 9).

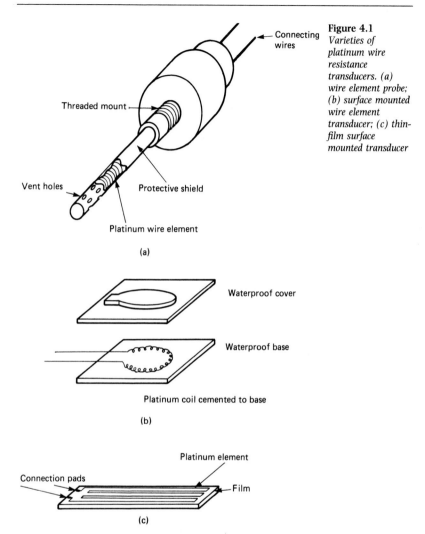

Figure 4.1
Varieties of platinum wire resistance transducers. (a) wire element probe; (b) surface mounted wire element transducer; (c) thin-film surface mounted transducer

Connecting wires

Threaded mount

Vent holes

Protective shield

Platinum wire element

(a)

Waterproof cover

Waterproof base

Platinum coil cemented to base

(b)

Platinum element

Connection pads

Film

(c)

Thermistors

A thermistor is essentially a semiconductor resistive device, whose resistance varies directly with temperature. The devices usually used for temperature measurement have a negative temperature coefficient (NTC) which means that the resistance falls with increasing temperature. Thermistors are typically used within the $-50\,°C$ to

Plate 4.2 *Sheathed element platinum resistance transducer* (RS Components)

+ 300 °C temperature range although some interface designs allow measurement outside of these limits. The main reason for this quite small temperature range is the thermistor's own non-linearity.

Semiconductor materials used to manufacture thermistors are generally sintered mixtures of sulphides or selenides, although oxides of cobalt, copper, iron, manganese or uranium are also used. The material is shaped into small beads, disks, rods or washers, which may then be encapsulated in glass, plastic or metal, or merely coated. Their small size ensures a reasonably fast dynamic response, and some miniature types are available which have a dynamic response of only a few microseconds.

Two main types of thermistor are found, either probe types or basic element sensors. Construction is essentially similar to basic resistors, except the resistive material used is, of course, temperature dependent. Thermistors are not usually close tolerance devices, which means that inter-

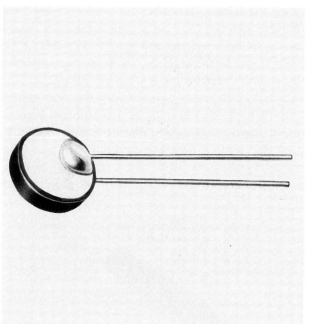

Plate 4.3 *Rod-type basic element thermistor* (RS Components)

facing circuits will need adjustment components to cater for differences in devices. However, some thermistors can have a tolerance of ± 0.2 ° C over their stated temperature range, which may be sufficiently precise to dispense with the need for adjustments.

Thermocouples

The sensing of temperature by thermocouples is based on the **Seebeck effect** (sometimes called the **thermoelectric effect**), in which two dissimilar materials are joined: if the two junctions so formed are at different temperatures, a current flows around the circuit.

The size of the current or the size of the EMF generated by the current is related to the difference in temperature between the two junctions and the materials used in the thermocouple construction.

In thermocouple measurement systems, the junction which is located at the temperature to be measured is called the **sensing junction**. The other junction (the **reference junction**) is usually maintained at a strictly defined reference temperature, such as ice point (0 °C). The reference temperature must be maintained to at least the accuracy required of the system. For the highest accuracy, in laboratory situations, say, the triple point of water is used. Alternatively, for less stringent requirements, a temperature-controlled oven can be used to maintain constant reference temperature. Cost of the necessary equipment to maintain the thermocouple reference temperature is a serious consideration in the design of

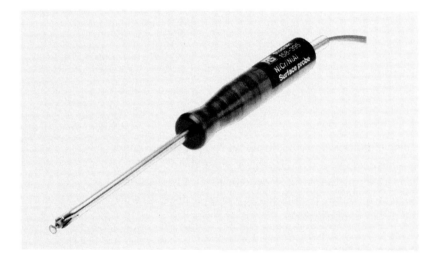

Plate 4.4 *Type K thermocouple surface probe* (RS Components)

temperature measurement systems. Temperature indicating interface systems are available, complete with automatic internal reference temperature compensation. Such thermometer systems display the sensor temperature to a reasonable degree of accuracy and eliminate the need for the user to calculate the sensor temperature from the EMF generated.

For greatest accuracy, tables exist for standard types of thermocouples, which relate the temperature difference (usually when the reference temperature is 0 °C) with the

Table 4.2 *Common thermocouple materials, with related standards. Types relate with the graphs shown in Figure 2.19 of thermal characteristics*

Type	BS number	First wire	Second wire
E		Nickel 90% Chromium 10%	Constantan (57% copper 43% nickel)
J	BS 1829	Iron	Constantan
K	BS 1827	Nickel 90% Chromium 10%	Nickel 94% Manganese 3% Aluminium 2% Silicon 1%
R	BS 1826	Platinum	Platinum 87% Rhodium 13%
S	BS 1826	Platinum	Platinum 90% Rhodium 10%
T	BS 1828	Copper	Constantan

EMF generated. Commonly used types of thermocouple, together with related British Standards and temperature ranges, are listed in Table 4.2.

Thermocouples are usually in probe form but as a thermocouple sensor merely requires the joining of two materials it is possible to have a thermocouple sensor which consists of two wires of the thermocouple materials which are joined at a fine tip. A number of different junction formations, made by welding, brazing, or silver soldering, are shown in Figure 4.2. The tip can then be

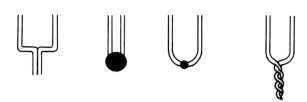

Figure 4.2 *Typical junction formations of thermocouple temperature transducers. From left to right: lap-welded; beaded; butt-welded; twisted wire*

inserted into a probe or can be directly inserted into the environment to be measured. Junctions can be grounded or ungrounded to a protective casing, and may be exposed or enclosed, as shown in Figure 4.3. A third form of

Figure 4.3 *Possible thermocouple probe tip types. From left to right: exposed and ungrounded; exposed and grounded; enclosed and grounded; enclosed and ungrounded*

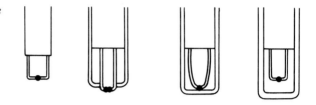

thermocouple consists of a junction mounted on a foil of some description, for surface mounting purposes. Typical construction is shown in Figure 4.4. The foil may be a magnetic strip, which is then easily attached to any ferromagnetic surface whose temperature is to be measured.

Figure 4.4 *Typical construction of a foil thermocouple transducer. The plastic film may be selfadhesive or magnetic, to aid mounting of the device*

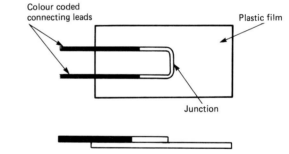

Thermocouples are rugged and, bearing in mind the requirements of a reference temperature, are quite economical. They have a reasonably fast dynamic response as they are typically small transducers, and they may be used over wide temperature ranges.

Other thermometers

In recent years, a number of other methods have arisen to sense temperature by thermometric means. Semiconductor temperature sensors, in integrated circuit

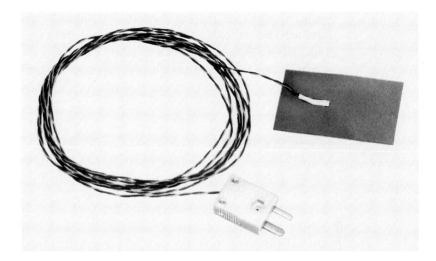

form, are available which produce an output current proportional to absolute temperature. Such transducers are reasonably linear in operation, and have sensitivities typically of the order of 1 μA of output current per degree kelvin. Connection into circuit is much as that of a thermistor, but more effective in remote temperature sensing systems, in that the current generated by the sensor is constant for any particular temperature so the resistance of connecting leads and consequent voltage drops becomes irrelevant.

Plate **4.5** *Type K thermocouple junction mounted on a flexible magnetic strip* (RS Components)

Other types of semiconductor temperature sensors produce a voltage which is proportional to temperature. One such device has an output voltage of 10 mV per degree kelvin.

Semiconductor sensors are, of course, based on the principle that the junction voltage and current of a PN junction varies proportionally with temperature. To this end, simple semiconductor diodes of germanium, silicon, gallium-arsenide or other materials, have all been used in temperature sensing applications as forms of temperature transducers. The negative temperature coefficient of the PN junction means that the voltage across the junction falls by about 2 mV per degree kelvin. Transistor characteristics change, too, with temperature and also allow integral amplification. It is the leakage current of the semiconductor junction which limits the upper range of semiconductor transducers, and ranges of $-50\,°C$ to $+150\,°C$ are typical for all such devices.

Radiation pyrometry

All the types of temperature transducer seen so far are thermometrical transducers, i.e., they rely on direct contact with the body whose temperature is to be measured. Radiation pyrometry is the science of measuring the temperature of a body without actual contact – it is the radiant energy from the body which is detected and thus gives the indication of the body's temperature. A **pyrometer** is a temperature transducer that detects radiant energy from a target: due to its very nature, therefore, a pyrometer is a remote (i.e., non-contacting) sensing transducer.

Plate 4.6 *Radiation pyrometer, with semiconductor infrared sensing element and integral buffer circuit* (RS Components)

In general a pyrometer transducer is a complete system, comprising:

1 An optical lens arrangement to focus the radiant energy onto the transduction element.

2 A transduction element to sense the amount of radiant energy.

3 Electronic circuits to interface the transduction element with following sub-systems.

All pyrometers rely on the fact that heat is emitted from a body by radiation. The principle is shown in Figure 4.5, where the pyrometer is represented by a box with an opening through which radiant energy from the target body passes to the transduction element. The amount of energy given off by the target body depends on its

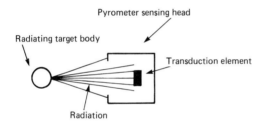

Pyrometer sensing head

Radiating target body

Transduction element

Radiation

Figure 4.5
Principle of a radiation pyrometer. The amount of radiation emitted by the target body depends on its temperature and its emissivity. Provided that the target completely fills the area 'viewed' by the pyrometer, and the target's emissivity is known, then the pyrometer output is a direct indication of the target's temperature

temperature and its **emissivity**. The emissivity of a body is a constant which depends on the body material itself. At first sight, it would appear that the amount of radiation which is picked up by the pyrometer depends on the distance from the target: the amount of radiation received is inversely proportional to the square of the distance between the pyrometer and the target (Figure 4.6(a)). Put another way, the amount of radiation reaching the transduction element of the pyrometer decreases proportionally with the square of the distance from the target. However, the area of the target 'viewed' by the transduction element in the pyrometer *increases* proportionally with the square of the distance from the target (Figure 4.6(b)). So, the radiation reaching the pyrometer does not depend upon its

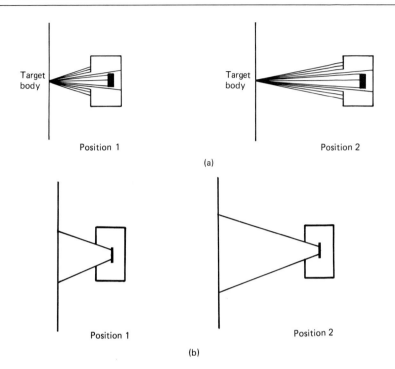

Figure 4.6
Showing how pyrometer's distance from the target body makes no difference to the pyrometer output: (a) although the amount of radiation from one spot on the target body decreases with distance; (b) the area 'viewed' by the pyrometer similarly increases with distance

distance from the target (with the proviso that the target completely fills the area 'viewed' by the pyrometer). It follows that if the emissivity of the target is known, then the output of the pyrometer is a direct indication of the temperature of the target.

The transduction element of a pyrometer may be formed by any of the temperature transduction methods already seen. Common pyrometers use thermocouple elements, resistance elements (pyrometers of this type are known as **bolometers**), or semiconductor elements. Although pyrometers of these types are typically large devices, recently some pyrometers have become available with elements measuring the radiation directly using a photo-electric or pyroelectric effect, in an integrated circuit form. These **pyroelectric detectors** still maintain the three basic parts of a pyrometer but are miniature, cheap, rugged and interface directly into the measurement system.

Pyroelectric nature in these transducers is dependent on a ceramic slice: heated during manufacture in the presence of an electric field, aligning the crystal dipoles. As the material cools the alignment is more or less retained, effectively forming a capacitor. The temperature of the slice defines the exact amount of alignment, and so the charge across the capacitance also varies with temperature. Changeable filters positioned across the radiation window of the pyroelectric detector allow the device to be made sensitive to different parts of the radiation spectrum, and an integral field effect transistor allows amplification and interfacing to subsequent circuits in the measurement system.

Because pyroelectric detectors sense temperature remotely, i.e., without physical contact, they may be used to detect the presence of a body, forming the basis of a **presence** or **proximity detector**.

Thermostats

Thermostats are a particular type of temperature transducer which open or close one or more electrical contact when a defined temperature point is passed. They provide a discrete temperature sensing method, rather than a continuous one. Because of this, they are sometimes called **temperature switches** and are rarely used in measurement systems. Often, on the other hand, they are used in control applications where all that is required is an indication that temperature has risen above, or fallen below, the desired point. A room thermostat is a good example.

Most common thermostats comprise a bimetallic sensing element, in which two metals having dissimilar thermal coefficients of expansion are mechanically joined in a strip. As temperature varies so the lengths of each individual layer of metal vary differently, thus the strip attempts to bend, eventually snapping into a concave shape, and closing electrical contacts. Bimetallic element thermostats are available in a number of forms.

Semiconductor thermostats, which exhibit similar

properties to bimetallic forms are available. Totally solid-state, i.e., with no moving parts, these devices act merely as a high resistance (approximately 100 kΩ) when an ambient temperature below the switchover point is encountered. However, at temperatures above the switchover point resistance is dramatically lowered (to about 100 Ω). Thermostats are available according to a wide variety of nominal switchover points (quoted in degrees Celsius), although in practice switchover takes place over about ± 5 °C of the nominal temperature.

5

Solid measurands

Measurands covered in this chapter, along with associated transducers, are those measurands of quantities relating to the mechanics of solid bodies. These are, in alphabetical order: acceleration, displacement, force, mass, motion, position, speed, strain, stress, torque, velocity, vibration, weight, although measurands will be covered in a non-alphabetical way.

Strain, stress

The most common transducer used in the detection and measurement of strain is the **strain gauge**, comprising a thin conductor or semiconductor element, mounted on the surface on which the stress or strain is to be exerted. As the surface itself undergoes small elongations or contractions due to the forces upon it, so does the strain gauge element, resulting in a change in gauge resistance due to the piezoresistive effect.

Plate 5.1 *Single polyester backed foil strain gauge (RS Components)*

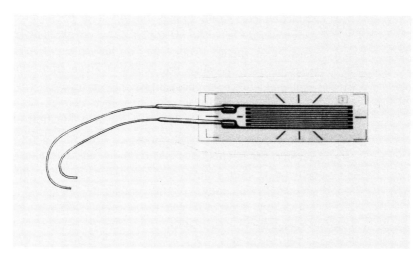

Although the strain gauge is purely a detector of the strain occurring on the surface where it is positioned, the strain may be caused by a number of measurands. So, in turn, the strain gauge can effectively detect all those measurands – typically: pressure, weight, mass, stress, torque, etc.

The two main parameters of strain gauges are **gauge factor** and resistance. Gauge factor is the sensitivity of the gauge element: the ratio of the percentage change in resistance to the percentage change in length.

A number of types of strain gauge is available. The simplest is the single gauge, polyester backed foil type; in which a metal foil is photo-etched to produce the gauge element. Extremely small sizes are available (down to less than 1 mm), and are quite stable in extremes of temperature and loading.

Transducers with two or more gauge elements are also common and are available in a variety of shapes. Where two or more gauges are mounted the configuration is known as a **rosette**, and can be used to measure magnitude and direction of more than one strain under complex loading conditions.

Gauge factors of metal foil gauges are around 2 to 4, and they have resistances commonly between 100 to 1000 Ω. Semiconductor strain gauges, on the other hand, have gauge factors between 50 to 200 and are thus more sensitive to applied strain: often no amplifying circuits need be used because the output voltage from the semiconductor strain gauge bridge is in the order of volts. Unfortunately, however, the resistance change of a semiconductor strain gauge element as a function of applied strain is quite nonlinear over its total strain range. Further, the resistance of a semiconductor strain gauge element is quite temperature dependent. So, although metal foil strain gauges require amplification, linearity is not usually a problem and temperature effects can be compensated for (see later).

Two main types of semiconductor strain gauge are

found: diffused or bare, where the semiconductor element is mounted directly onto the surface to be measured; and encapsulated, where the element is mounted on a carrier which in turn is mounted on the surface to be measured. Bare gauges offer advantages in terms of physical size (as little as only a few square millimetres) and accuracy, but they are also very fragile, requiring careful handling and mounting. Like foil gauges, dual-element and multi-element rosette semiconductor devices are available.

Displacement, position and motion

Strain gauges operate because the overall element length is changed upon application of a strain. Put another way, if two points on the element are considered to have been *displaced* then strain has occurred. So, the strain gauge can effectively measure displacement, albeit displacement of only a very small amount. To this end, strain gauge displacement transducers are available, usually comprising a bending beam or flexure to which are attached strain gauge elements. These are not common as such, however, but are more usually found as part of transducers designed to measure other measurands (see relevant sections on weight, acceleration, pressure, etc.).

Encoders

As far as quite large displacements are concerned (over a metre or so), one of the biggest groups of displacement transducers is formed by encoders. These are transducers which are capable of giving an output which is digitally coded according to the sensed displacement. Linear displacement is sensed by **linear encoders** and rotary displacement is sensed by **rotary** or **angular encoders**. Three main transduction principles are used to make encoders, illustrated in Figure 5.1. Essentially, all types have a moving disc or strip of some description, comprising a number of tracks which are scanned by: brushes (Figure 5.1(a)); light beams (Figure 5.1(b)); or magnetic coils (Figure 5.1(c)).

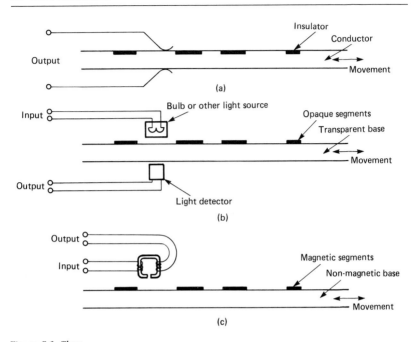

Figure 5.1 *Three main transduction principles of encoders. (a) brush type in which brushes make or break contacts through the conductive base of disc or strip; (b) optical type; in which the base is transparent and opaque segments cut off the light between source and detector – alternatively a reflective base with nonreflective segments is used; (c) magnetic type; in which a twin coil is used to detect whether a magnetic segment or the nonmagnetic disc or strip base is beneath it*

In the brush type of encoder, the disc or strip is conductive but has insulating segments on its surface. When the brush is in contact with the base the circuit is closed; when in contact with an insulating segment the circuit is open. For fairly obvious reasons of contamination, friction, limited life, etc., brush type encoders are not usually the ideal choice.

In the optical encoder, the disc or strip comprises either a transparent layer with opaque segments on its surface, or a reflective layer with non-reflective segments. In the former, when an opaque segment is between the light source and sensor the light beam is interrupted, when no opaque segment lies between the source and sensor the beam is complete. In the latter, light is either reflected or not reflected between the light source and sensor, completing or interrupting the beam.

In the magnetic encoder, the disc or strip is non-magnetic but has on it a pattern of magnetized segments, and a pickup head with input and output coils. An

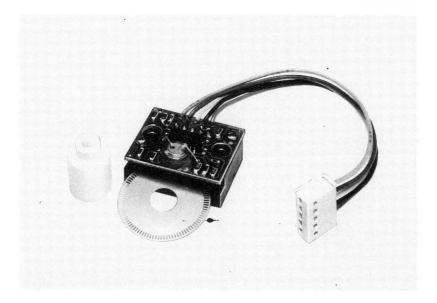

interrogate signal (typically a 200 kHz constant amplitude pulse, or a positive-going logic pulse) is applied to the input coil, so that when the head is over the base an output is obtained from the output coil. However, when the head is over a magnetized segment, no output occurs.

Plate 5.2 *Optical incremental angular encoder* (RS Components)

Two main forms of encoder are available: the **incremental encoder** and the **absolute encoder**. The names refer to the facts that incremental encoders can only indicate displacement as a number of increments since the movement started; whereas absolute encoders indicate the absolute position. Put another way, the incremental encoder must be returned to the starting position whenever the system is first turned on – so that the system has a datum point; the absolute encoder does not – its output is a direct statement of position.

Incremental encoders (Figure 5.2) produce output pulses which are counted by an up/down counter, so the value of the count is an indication of how far the disc or strip has moved since the count began. Often, two sensing elements are used, positioned on the transducer so that

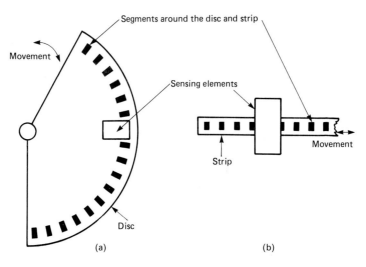

Segments around the disc and strip

Movement

Sensing elements

Strip

Movement

Disc

(a)   (b)

Figure 5.2
Principle behind incremental encoders: (a) angular encoder; as the disc rotates the sensing element detects movement as a number of segments pass beneath; (b) linear encoder; the strip moves underneath the sensing element so allowing movement to be detected

their outputs are 90° out of phase; enabling simple logic circuits to determine direction, and hence whether the counter must count up or down.

Resolution of incremental encoders of this type depends directly on the number of sensing segments on the disc or strip within the transducer. Mechanically, therefore, there is a limit to the resolution available. **Optical interference incremental encoders** substantially improve resolution, by effectively magnifying the distance between sensed segments. Moiré fringe patterns are set up by having two plates, each with a series of parallel opaque strips, positioned so that the strips are at a small angle (Figure 5.3). If one plate moves relative to the other in a direction perpendicular to the strips Moiré fringes occur, with the appearance of a dark strip moving in the direction of the strips. The fringes move up if relative movement between the plates is in one direction, and down if movement is in the other direction. Furthermore, the length of fringe movement is much more than the distance between strips so a 'magnification' of the apparent resolution takes place, where the magnification factor approximates to the inverse of the angle in radians between the strips (for small angles, only). So, for an incremental encoder with strips

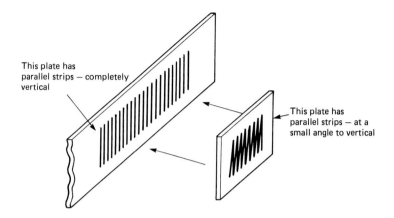

This plate has parallel strips — completely vertical

This plate has parallel strips — at a small angle to vertical

spaced 0.05 mm apart, where the two plates are at an angle of only 0.01 radians, the magnification factor is 100 and the fringes are 5 mm apart.

Absolute encoders (Figure 5.4) produce a coded output which indicates *absolute* position by the very fact that it *is* a code. Ideally, at least in terms of ease of use, the code used would be the same binary digital code used by all microprocessors and computing equipment, with a bit-length which matches whatever bit-length is used by the measurement system. For an angular encoder with a bit-length of 20 bits, transducer resolution affords the prospect of being able to measure to one millionth of a

Figure 5.3
Production of Moire fringes in an optical interference incremental encoder. As the two overlapping sets of parallel strips are at a small angle, light passing through the two plates creates a set of Moire fringes, allowing movement to be detected more easily

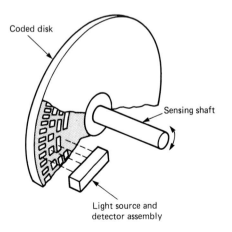

Coded disk

Sensing shaft

Light source and detector assembly

Figure 5.4
Principle of a rotary absolute encoder. The disc is marked with a code which allows the sensing element to detect the exact rotary position of the disk. Same principle is used in linear absolute encoders

revolution! However, the conventional binary code causes significant measuring problems because it is perfectly possible for two adjacent codes to require more than one individual bit to change state between the two. For example, a change of state from the code 0011 to the code 0100 (a single step between the binary digitally coded numbers three and four) theoretically requires that three different bits simultaneously change state. Practically, though, it is impossible to ensure that all bits which are to change state between two adjacent codes do so simultaneously. In this example, therefore, codes 0010, 0001, 0000, 0111, 0110 and 0101 could all occur as the codes change.

The simplest way to avoid this problem is to use a digital sequence of codes in which only one bit is required to change state between each position. An example is the **Gray code**, which for a four-bit arrangement is:

0000
0001
0011
0010
0110
0111
0101
0100
1100
1101
1111
1110
1010
1011
1001
1000

Other sequences are used, too, but the Gray code is the most popular. Whatever code is used, other than the conventional binary digital version, the measuring system must initially convert the output from the encoder.

Reading ambiguities on an encoder using conventional binary code may also be minimized using a number of scanning techniques, which typically utilize two transduction elements per track, one slightly ahead (leading) and the other slightly behind (lagging), together with external logic circuits to determine the actual output. Most common of these techniques is the V-scan, but others such as the U-scan and the M-scan are also used.

The principle of all scanning techniques is that whenever a conventional binary sequence is changing from one state to an adjacent state, the LSB must change state. Further, for an increasing binary number, if the LSB changes from 0 to 1, no other bits change state. If the LSB changes from 1 to 0, however, at least one other bit must change state also. For a decreasing binary number the reverse is true. The external logic circuits have to interpret these occurrences and switch between leading and lagging transduction elements as required to ensure nonambiguity of output code.

One disadvantage of any binary digital code, is that the total number of codes in the sequence is equal to a power of two (for example, in the four-bit Gray code shown here, the total number of codes is $2^4 = 16$). In the case of a rotary encoder, however, it may be required that the output is a direct indication of angular position in degrees. As there are $360°$ in one revolution, it is impossible to derive a binary digital *cyclic* code such as the Gray code with 360 codes. In such an encoder, it is usual to electronically mask the output during the transition from $359°$ to $0°$, producing an all-zero code which does not suffer from the ambiguity problem, because all the zeros are generated from the same signal and so must occur simultaneously.

Potentiometric displacement transducers

An extremely simple form of displacement transducer can be made with a resistive potential divider, in which a sliding contact or wiper moves over a resistance element (Figure 5.5). The wiper is mechanically connected with

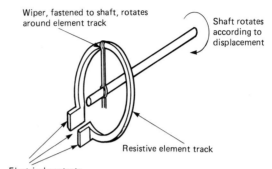

the sensing shaft of the transducer which moves according
to the displacement to be measured. If a voltage is applied
across the resistive element, the voltage at the wiper is
indicative of the displacement.

Rotary and linear potential divider displacement transducers are common; rotary types are available with ranges
as little as just a few degrees or as much as 7200° (i.e., 20-
turn); linear types have ranges between just a few
millimetres up to several metres. Both types require
electrical connections to the resistance element, which
takes up space and, possibly, limits rotary travel to less
than might be otherwise assumed (for example, a single-
turn transducer may be usable over only about 350° of
travel).

The form of the resistance element has a bearing on the
transducer's resolution. If the element is of a wirewound
form, resolution steps are given by the number of turns of
wire per length of element. A number of other potential
divider transducers use films of plastic, metal, carbon, or a
ceramic-metal mixture (known as cermet) to create the
resistance element and thus give, theoretically, infinite
resolution.

Capacitive displacement transducers

A number of types of capacitive displacement transducers
are available; all working on the principle that a capacitor
is made with two plates separated by a dielectric. Varying

Plate 5.3 *Linear potential divider displacement transducer. Sensing element is formed with a conductive polymer resistance* (RS Components)

Plate 5.4 *Precision linear potential divider displacement transducer* (RS Components)

Plate 5.5 *Rotary
potential divider
displacement
transducer. Spring-
loaded to return to
zero position, with
105° of revolution
(RS Components)*

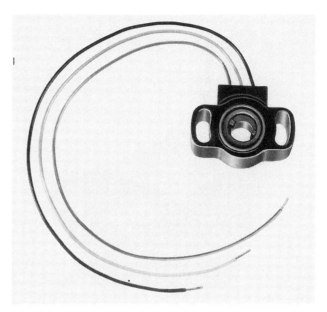

Plate 5.5 *Rotary potential divider displacement transducer. Spring-loaded to return to zero position, with 105° of revolution (RS Components)*

the plates, the separation, or the dielectric causes a change in capacitance.

Figure 5.6 shows the basis of capacitive displacement transducers, where the capacitance is varied by: moving one plate in relation to the other (Figure 5.6(a)); altering the area of plate overlap (Figure 5.6(b)), or moving the dielectric (Figure 5.6(c)).

There is also the possibility that the dielectric constant itself may be altered: this principle lies behind the use of **capacitive proximity sensors** and **capacitive proximity switches** capable of detecting targets at a distance from the device, and the **capacitive fluid level transducer**, simply because the target's presence or a changing level of fluid alters the dielectric constant.

Inductive displacement transducers

The self-inductance of a coil varies with the proximity of a magnetically permeable body, so the displacement of the body from the coil can be measured by sensing the coil's self-inductance. Transducers using this principle are

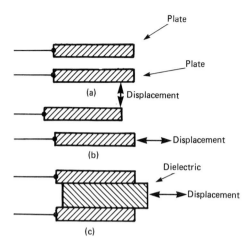

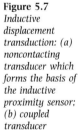

Figure 5.6
Capacitive displacement transducer: (a) as the plates move apart or together with the displacement, so the capacitance changes; (b) as the plates move with the displacement, so the overlapping surface changes, changing the capacitance; (c) as the dielectric moves with the displacement so the capacitance changes

usually noncontacting (Figure 5.7(a)), with the displacement of the body influencing self-inductance merely by proximity, but coupled inductive displacement transducers (Figure 5.7(b)) do exist, in which the coil's core is mechanically coupled to the device whose displacement is to be measured. Noncontacting transducers are most commonly encountered as the basis of **inductive proximity sensors** or **inductive proximity switches**.

Reluctive displacement transducers
Transducers in this important category all function on the principle that the reluctance path between two or more AC excited coils is varied with the displacement to be measured, causing an output AC voltage change. The transduction elements so formed are used in many other

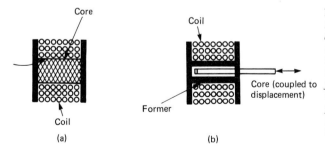

Figure 5.7
Inductive displacement transduction: (a) noncontacting transducer which forms the basis of the inductive proximity sensor; (b) coupled transducer

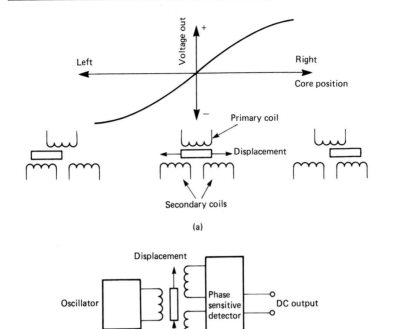

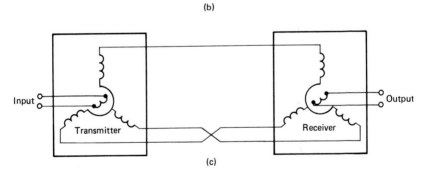

Figure 5.8
*Reluctive
displacement
transducers: (a) the
linear variable
displacement
transformer (LVDT)
whose output
voltage varies in
phase and amplitude
according to core
position; (b) using a
phase sensitive
detector with the
LVDT to determine
core position; (c)
synchro; shown in
its usual mode or
operation as a
synchro transmitter
and synchro
receiver; (d) resolver*

categories of transducers, in which the sensing elements convert measurand changes into displacements, which can then be measured and thus indicate measurand value. Operation of some of the many different types of reluctive displacement transducer is shown in Figure 5.8 and described below:

- **Linear variable differential transformer** (LVDT): the three coils are wound in the same axis, the centre coil being the primary. With AC excitation to the primary, when the core moves within the coil assembly the coupling between the primary and secondaries changes, resulting in output voltage magnitude and phase changes at the secondary terminals. Figure 5.8(a) shows output voltage and phase for an LVDT, showing that zero voltage is obtained when the core is in the central or null position. Output changes are, in fact, nonlinear, but are usually considered linear (and can have linearity errors of less than 1%) over the central region of core travel. Displacements of only a few millimetres up to about a metre can be measured with LVDTs. Operation of the **rotary variable differential transformer** (RVDT) is similar, although angular displacement is usually limited to about ± 40°

  Unfortunately, one problem when using the basic differential transformer is the very AC output voltage it relies on, together with the associated phase changes. Because of this, LVDTs and RVDTs are usually used with a phase-sensitive detector (as in Figure 5.8(b)) which provides a DC output voltage; positive when the core is at one side of the null position, negative at the other side, and zero when the core is at null. Many differential transformer transducers are available with internal oscillator, detector, filter and amplifying circuits which allow transducer use with a DC excitation voltage, giving a smoothed DC output, too. These DC–DC devices provide the user with a useful transducer which is extremely easy to use.

● **Synchro**: the single primary rotor winding is turned with the rotary displacement to be measured. Three stator secondary windings, at 120° spacings, form the output coils. AC excitation to the rotor, normally called the **reference frequency**, induces outputs from the three secondaries and the relationship between the outputs is a direct consequence of rotor position.

Typically, synchro displacement measuring systems comprise two similar units, a **synchro transmitter** (sometimes called a **control transmitter**) and a **synchro receiver** (sometimes called **synchro transformer** or **control transformer**). The two synchros are connected as in Figure 5.8(c), such that displacement of the transmitter's rotor causes the receiver's rotor to turn to the same position – the receiver can therefore be used as the basis of a display device, its rotor position indicating the rotary displacement being measured, or can be used to do other mechanical work. In this mode of operation the synchro transmitter can also be known as a **torque synchro**. Used alone to provide control signals for other measurement systems, a synchro transmitter is known as a **control synchro**.

The synchro principle is also used in the **inductosyn**, a linear displacement transducer, in which a flat stator secondary winding comprising a pattern of conductors forms a linear 'scale', over which two 'scanning head' primary windings traverse.

● **Resolver** (Figure 5.8(d)): basically a synchro with only two stator windings (this time primary windings, and at 90°) but two rotor windings (this time secondary windings, also at 90°). When used purely as a displacement transducer, providing output signals to a measurement system, one rotor winding is shorted.

Switches and proximity devices
Contacting and noncontacting switches can be effectively used as position sensing transducers; the simplest being the microswitch which operates by a small, but neverthe-

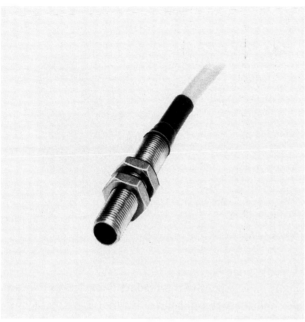

Plate 5.6 *Inductive proximity switch, with threaded barrel to allow accurate position of sensing head* (RS Components)

less, physical connection between it and the body being sensed. Proximity devices, on the other hand, have no physical connection, the presence of the body being sensed by its affect on some principle of the transducer's operation.

Inductive proximity sensors are available, in which a tuned oscillation is maintained. When a conductive body approaches the sensor, damping of the oscillation occurs, which is sensed by interfacing circuits. Common too, and much more easily used, are inductive proximity **switches**, which house interfacing circuits so that presence of a conductive body produces a definite switched on/off action.

Capacitive proximity sensors also fall into sensor or switch categories, generally operating on the principle that a capacitor in a balanced bridge circuit causes the bridge to become unbalanced if a local body changes the capacitor's permittivity. Capacitive proximity devices are

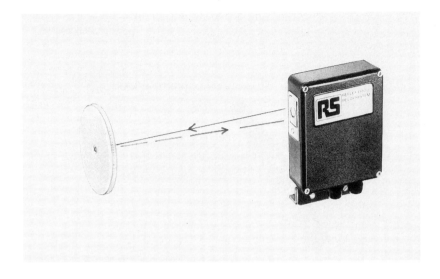

Plate 5.7
*Retroreflective scan
optical proximity
transducer, with
reflective disc* (RS
Components)

usually more expensive than inductive devices as they are more difficult to manufacture, but they have the advantage of being able to sense presence of a wide range of body material, over greater distances.

A large number of **optical proximity devices** is available, in both basic sensor and integral interface circuit forms. Two operating modes are common: **direct** or **through scan**, and **reflective scan**, illustrated in Figure 5.9. Reflective scan devices can be either **retroreflective** (Figure 5.9(a)) in which the light source and sensor are commonly housed and the incident light is reflected back to the sensor along its pre-reflected path; **specular** (Figure 5.9(b)) in which the beam of light between source and sensor is reflected at an angle off a mirrored or, at least, mirror-like surface, or; **diffuse** in which the reflecting surface is of a matt nature.

Magnetic proximity transducers, containing a magnetically operated switch of either reed relay or Hall effect types are also commonly used for proximity sensing.

Radar proximity transducers are available, which comprise some form of radar frequency oscillator and mixing device to combine the reflected and generated signals.

Doppler effect, in which moving bodies produce a difference in frequency between the two signals, means that
output frequency of the mixing device is 0 Hz with no
moving body, but above this if a moving body is present.
The most common form of this type of transducer operates
at X band microwave frequencies.

Use of such a device is not restricted to proximity
sensing; distance and, indeed, speed of bodies can be
measured by measuring the time a transmitted pulse takes
to return after reflection and performing required calculations on these measurements.

Acceleration and vibration

The principle behind operation of all acceleration transducers (often called **accelerometers**) is that the acceleration which the transducer is subjected to forces a mass
(called the **seismic mass**) to move within the transducer.
The seismic mass is attached to a spring or similar device
which counteracts the movement of the mass, such that
the mass moves until the force from the spring balances
the force the mass experiences due to the acceleration.

Figure 5.9.
*Principles of optical
proximity devices.
Reflective devices
can be: (a)
retroreflective; (b)
specular. Direct
device principle is
shown in (c)*

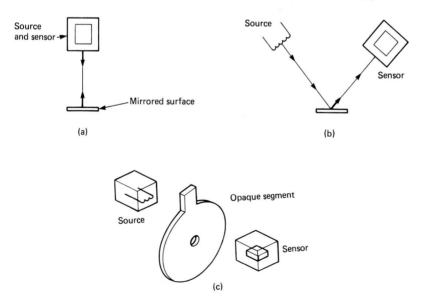

Plate 5.8
Retroreflective scan optical proximity transducer, comprising infra-red LED source and phototransistor sensor mounted in miniature housing (RS Components)

Measurement of the seismic mass's movement in relation to the body of the transducer allows the acceleration itself to be calculated. Figure 5.10 illustrates the principle with a simple seismic mass and spring arrangement, together with a measurement scale within the transducer. In the unmoved state (Figure 5.10(a)), the mass is stationary and the scale displays zero, but if an accelerating force is applied to the transducer, the mass will tend to stay in its original position due to its inertia while the transducer body moves. The result is a change in scale position, indicating the size of the acceleration.

Most accelerometers are not quite as simple as this, of course, although some do exist. One of the main disadvantages of this setup is that the transducer can only measure acceleration in the same axis as the spring-mass-scale arrangement. Often acceleration may be in more than one axis, and so transducers capable of measuring in two dimensions (biaxial) or three dimensions (triaxial) are available.

A practical problem is that of damping. If an acceleration is applied to the simple arrangement we have just seen as a step function, then the mass will oscillate around the

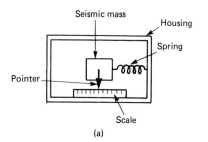

Seismic mass

Housing

Spring

Pointer

Scale

(a)

Seismic mass tends to remain in position while housing moves

Acceleration

(b)

final output; the oscillation taking a length of time to decay. Only if a certain level of damping is applied to the mass will the reading be accurately available after an adequately short period of time. The situation arises because accelerometers, because of the measurand they are attempting to measure, are second-order transducers, i.e., acceleration is represented mathematically by a second-order differential equation of the form:

$$a = \frac{d^2x}{dt^2} + 2b\omega_n\frac{dx}{dt} + \omega_n^2x$$

See Chapter 3 for a more detailed explanation of second-order transducers.

Figure 5.10 *Accelerometer principle. An applied acceleration moves the housing, but the seismic mass tends to remain in its original position due to inertia*

Plate 5.9 *Optical proximity switch which can operate in retroreflective or direct scan modes with optical fibres (RS Components)*

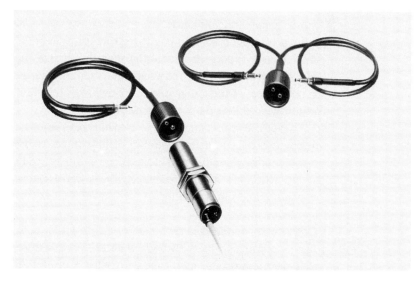

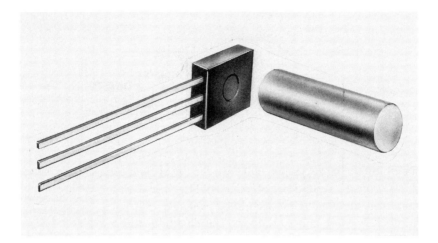

Plate 5.10 *Hall effect magnetic proximity transducer* (RS Components)

Damping is provided in many types of accelerometers by enclosing the arrangement in a chamber and filling the chamber with silicone oil. Other methods of damping are oil-filled dashpots, electromagnetic damping or simple air damping arrangements.

Although all the transduction principles are employed in the various available accelerometers, two principles are most common: strain gauge and piezoelectric.

Strain gauge accelerometers

If the acceleration acting upon the spring-mass arrangement causes the movement of the mass to change the dimensions of one or more strain gauges, then the change in resistances of the gauge or gauges owing to resultant strain is an indication of the acceleration.

All types of strain gauge can be used in accelerometers. Readers are referred to the beginning of this chapter, to the section on the measurement of stress and strain by strain gauge techniques, for a more detailed discussion of the use of strain gauges in transducers.

Piezoelectric accelerometers

The seismic mass of piezoelectric accelerometers acts upon piezoelectric crystals, such that an applied acceleration

causes the crystal to be tensed or compressed, changing the electric charge across the crystal. Typical makeup is shown in Figure 5.11, where a circular crystal element is bolted with an insulating bolt and washers, together with the seismic mass, to the transducer housing. Electrical connections are made to the piezoelectric element via metal layers. Note that a separate spring arrangement is

Plate 5.11 *Radar proximity transducer, operating at microwave frequencies (RS Components)*

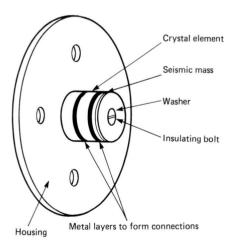

Figure 5.11 *One method of mounting a piezoelectric crystal element as an accelerometer*

Crystal element

Seismic mass

Washer

Insulating bolt

Housing   Metal layers to form connections

not required, as the bolt is tightened to preload the element. The system itself therefore provides the 'spring'. Another piezoelectric element mounting method (Figure 5.12) has a central pillar through the piezoelectric element which acts as the seismic mass. The element is bonded to the pillar, so that the element distorts as the pillar moves upon application of acceleration. In this type of accelerometer, the element is the 'spring'. This is an

Figure 5.12
*Another method
mounting a
piezoelectric crystal
element as an
accelerometer*

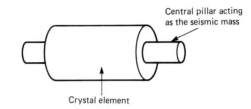

Central pillar acting
as the seismic mass

Crystal element

advantage over the previous type of piezoelectric accelerometer, in terms of weight, because no account need be taken of mounting effects on the transducer's operation. The system of Figure 5.11, on the other hand, must have a sufficiently strong (and thus heavy) housing to prevent mounting effects on the housing affecting the element via the loading bolt.

The high output impedance of piezoelectric crystal elements means that interfacing leads are highly susceptible to interference. To counteract this problem piezoelectric accelerometers are available with an integral amplifier acting as an impedance converter.

Force, mass
Transducers designed to measure force are very often used in weighing applications, and so are sometimes called **load cells**. Generally, whatever force is to be measured, whether it is the force of some unknown mass due to gravity or any other force, for example, pressure, torque, an indirect measurement is taken. Typically, the applied force is converted into a mechanical displacement of an elastic sensing element: the displacement of this element is then

Plate 5.12 *Strain gauge load cells with beam-type force sensing elements* (RS Components)

measured with the use of some transduction element. Most of the basic transduction principles are used in the many types of force transducers, although it is fairly obvious that the most common principle is going to be that of piezoresistance – in the strain gauge. Readers should, therefore, refer to the section on strain and stress measurement, discussed earlier in this chapter.

Sensing elements

Four types of sensing elements are used in force transducers; beams, cylinders, diaphragms and proof rings. Figure 5.13 shows the principle of a beam-type force sensing element, where a beam is fixed either at both ends (Figure 5.13(a)) or at one end (Figure 5.13(b)) and the force is applied perpendicularly somewhere along its length. Deflection of the beam is a measure of the applied force magnitude.

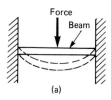

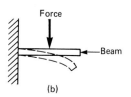

Figure 5.13 A beam-type force sensing element. Applied force in any direction causes the beam to bend. Beam can be either: (a) mounted at both ends; (b) mounted at only one end

Figure 5.14 shows the principle of cylinder-type force sensing elements, in which the force is applied along the axis of the cylinder. The cylinder deflects outwards when force is applied, and the deflection is an indication of the magnitude of the force.

Figure 5.14 *A cylinder-type force sensing element. Applied force causes the cylinder to deflect outwards*

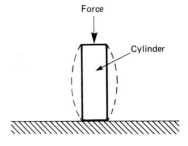

Diaphragm-type force sensing elements (Figure 5.15) comprise a circular plate, clamped around its edge. Function is essentially similar to the beam sensing element, in that the diaphragm deflects at its centre when a force is applied. Deflection is indicative of force magnitude. Diaphragm sensing elements are most commonly found in

Figure 5.15 *A diaphragm-type force sensing element. Applied force causes the element to deflect*

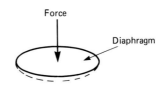

pressure transducers, where the applied force is the pressure of a gas, etc. For this reason, diaphragm-based transducers are covered in greater depth in the section on pressure transducers in Chapter 6.

Proof rings (Figure 5.16) can be of circular shape as shown or of square or rectangular shape (when they are usually called **proving frames**). The applied force distorts the shape of the ring; once again deflection being an indication of the force magnitude.

Speed, velocity
Transducers in this category are used to measure either

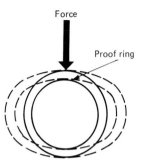

Force

Proof ring

linear or angular velocity. Angular velocity transducers are usually of electromagnetic means and are known as **tachometers**. Some electro-optical tachometers are available (**stroboscopes**). Linear velocity is most commonly sensed indirectly by converting linear velocity into a rotation via a wheel or gear system, then using an angular velocity transducer to measure the wheel or gear's speed. Direct sensing means are generally restricted to small movements in electromagnetic linear velocity transducers or remote-sensing Doppler shift microwave radar means.

The incremental encoder, previously discussed, can also be used as a velocity transducer: the rate of pulses corresponding to sensed segments being proportional to velocity.

Pulse tachometers

The most common method of transduction is one in which a pickup coil of some description allows detection of a rotating shaft. Generally the shaft is spiked or takes the appearance of a toothed gear (Figure 5.17). As a tooth or spike passes the pickup, output voltage changes to indicate

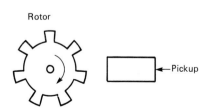

Rotor

Pickup

Figure 5.17 *Pulse tachometer principle: a rotor, with toothed protrusions is sensed in some way by a pickup. As each protrusion passes the pickup, ouput signal indicates the occurrence*

Plate 5.13 *Electro-magnetic tachometer transducer* (RS Components)

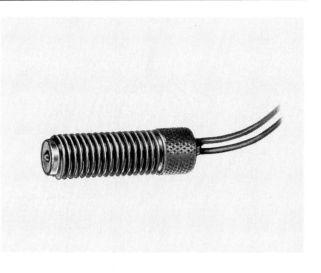

the occurrence. A count of the number of output voltage changes over a given period indicates the angular velocity.

A number of principles may be used in the transduction element of such transducers for example, Hall effect, inductive eddy current type, or even optical arrangements (proximity transducers), but by far the most common principle used as a pulse tachometer is the electromagnetic principle.

Plate 5.14 *Optical tachometer with integral light source and sensor, giving direct readout* (RS Components)

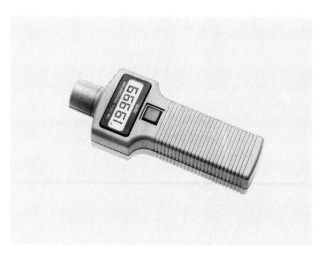

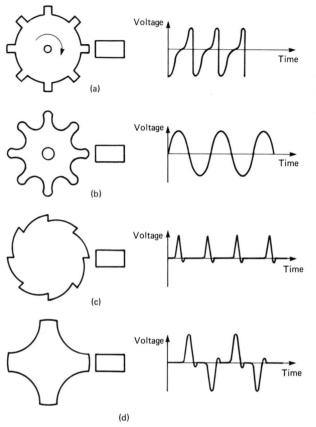

A ferromagnetic rotor is used with a pickup comprising a permanent magnet and coil. The magnet creates a magnetic field around the sensor. When a tooth on the rotor passes through the field, the flux change induces an EMF in the coil. One advantage of this method over others is that the pulsed output shape depends upon the rotor's tooth shape. A number of pulsed output forms are shown in Figure 5.18 for various tooth shapes.

Tachometer generators
If the velocity to be sensed can be coupled to a generator of some description, then the signal voltage or frequency at

the generator output can be used to indicate the velocity's magnitude. Generators used for this purpose are often called **tachogenerators**.

One problem with the tachogenerator technique is that power is required to drive the transducer. The velocity must have sufficient energy so that the power required to drive the tachogenerator is negligible and so does not affect the velocity magnitude.

Stroboscopes

Stroboscopes are a special form of electo-optical tachometers, using a series of light flashes to illuminate the rotating shaft. The flash rate is adjusted manually until the shaft appears to be stationary – which occurs when one flash takes place on each cycle of rotation.

6

Fluid measurands

This chapter looks in detail at the transduction of fluid mechanical measurands, i.e., those measurands relating to gas and liquid quantities. In alphabetical order these measurands are: flow, humidity, level, moisture, pressure.

Flow

The term *flow* means the motion of a fluid – either a liquid or a gas. *Flowmeters* – transducers used to measure flow actually indicate **flow rate**, the amount of the fluid which flows over a given period. There are three ways in which flow rate can be defined:

1 **Mass flow rate**, which is flow rate in units of mass of the fluid per unit time, for example $kg\ s^{-1}$.
2 **Volumetric flow rate**, which is flow rate in units of fluid volume per unit time, for example, m^3s^{-1}.
3 **Velocity of flow**, in units of $m\ s^{-1}$.

Most measurements of flow are concerned with volumetric flow rate, and in many cases mass flow rate can be calculated from this, although variations in pressure, density, temperature, etc., must be taken into account – particularly in the measurement of gas flow rates. Similarly, a measurement of velocity of flow can form the basis of a calculation to determine mass flow rate or volumetric flow rate.

Mechanical flow sensing

Flow sensing methods are varied. Simplest, and therefore probably most common, are the mechanical methods in which the flowing fluid displaces or rotates a solid body in such a way that displacement or rotation is proportional to

Plate 6.1 *Float-position sensing flowmeter. Liquid flows upwards through the flowmeter, causing an internal float to lift in relation to volumetric flow rate. Float position is sensed with a differential transformer arrangement* (Krohne Measurement & Control)

flow rate. Figure 6.1 shows the main methods used in such transducers. Figure 6.1(a) is the detail of a spring-loaded and hinged vane, which is pushed open as the fluid flows through the transducer. The greater the flow rate the further the vane is pushed open. Figure 6.1(b) shows a transducer which uses the same principle but with a spring-restrained plug. Many other variations of this principle are usable. Figure 6.1(c) shows the principle of a propellor which is rotated by the fluid, the rate of rotation is proportional to flow rate.

Most common of the mechanical flow rate transducers is the **turbine flowmeter**, which uses the rotating propellor

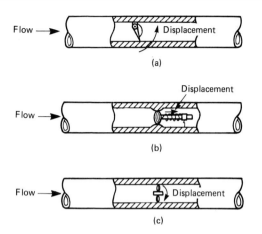

Figure 6.1 *Typical methods of mechanical flow sensing: (a) a spring-loaded hinged vane which is pushed open as fluid flows through the transducer; (b) a spring-restrained plug; (c) a propellor which rotates as fluid flows through*

(or, in this case, turbine). Figure 6.2 shows the main parts of a typical turbine flowmeter, in which the turbine is held by a bearing in the fluid flow. Generally, the blades of the turbine are ferromagnetic, so a coil attached to the flowmeter body can be used to detect turbine rotation. Electromagnetic sensing such as this creates a drag effect on the turbine which, at low rates, can affect turbine rotation. If a turbine flowmeter is to be used with low fluid rates some other pickup arrangement, say, electro-optical, is recommended.

For accurate measurements it is important that the fluid

Figure 6.2 *Turbine flowmeter, in which a turbine rotates as fluid flows through, and a coil is used to detect rotation of the ferromagnetic propellor blades*

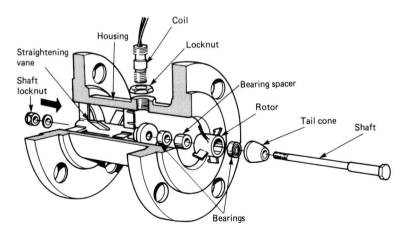

Plate 6.2 *Rotating propellor flowmeter, with electromagnetic sensing of propellor rotation* (Litre Meter)

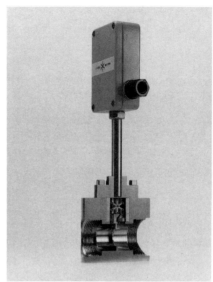

is not swirling (because this directly affects the speed of rotation of the turbine) so straightening blades are usually mounted at the entrance to the flowmeter. These blades also form one bearing point for the turbine. However, some much simpler turbine flowmeters are available if accuracy is not essential, i.e., if swirl and drag are not major considerations, in the measurement system.

Plate 6.3 *Electromagnetic flowmeter* (Krohne Measurement & Control)

One advantage of the turbine flowmeter over some other forms is that (over its stated range) the output signal is linearly proportional to flow rate of the fluid.

Differential pressure flow sensing

If a fluid is forced through some type of restriction in a pipe, its velocity changes causing a pressure difference which is proportional to flow. By measuring this pressure difference (usually using a differential pressure transducer) the flow rate can thus be detected.

The most common type of **differential pressure** or **variable pressure flowmeter** is the Venturi tube (Figure 6.3(a)), in which the fluid goes through a sort of 'bottleneck' in the tube. BS 1042 gives preferred sizes of Venturi tubes to measure a range of fluids.

Another common differential pressure flowmeter is the Pitot tube (Figure 6.3(b)), in which a probe tube is inserted through the wall of the main tube and is brought to face directly into the fluid flow. This probe is referred to as the **impact probe**. A second probe (the **static probe**) is taken to the tube wall. The difference in pressure between the impact pressure and the static pressure is an indication of flow rate. Other possible differential pressure flowmeters are illustrated in Figure 6.3(c) and (d). The principle is, however, similar.

One disadvantage of differential pressure flowmeters is the fact that the velocity of flow is proportional to the square root of the pressure difference, i.e., they are essentially nonlinear devices. Generally, too, they cannot be used for measurement of gas flow, because they rely on the fact that the fluid does not become compressed on passing the tube restriction. Gases are compressible, of course, so some correction must be taken into account.

Thermal flow sensing

Thermal flowmeters rely on the principle that the amount of heat transferred from one point to another in a moving fluid is proportional to the fluid's mass flow rate. Figure

Figure 6.3 *A number of differential pressure sensing transducers: (a) the Venturi tube, which allows measurement of the difference in pressure before and after a restriction in tube bore size; (b) the Pitot tube, which uses two probes to create a difference in pressure proportional to flow; (c) the orifice plate transducer; (d) the centrifugal section transducer*

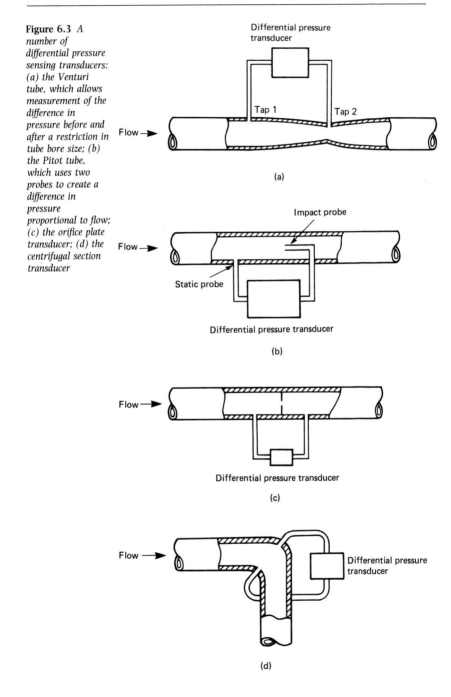

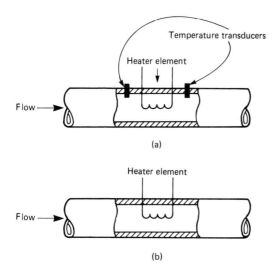

Figure 6.4 *Thermal flow sensing using a heater element: (a) with two temperature transducers; (b) by itself in the hot-wire anemometer*

6.4(a) illustrates how two temperature transducers measure fluid temperature before and after heating, and a heater element is placed between the two transducers.

Hot-wire anemometers take the principle one stage further, merely placing a single wire element in the fluid flow (Figure 6.4(b)). The cooling effect of the fluid flow as it passes the element is indicative of the mass flow rate – cooling is detected in the change in wire element resistance. Often, instead of a wire element, thin conductive films are used. Extremely fast fluctuations of flow rate can be detected by hot-wire anemometers.

Electromagnetic flow sensing
If a conductive fluid flows through a transverse magnetic field an electromotive force is induced in it, which is proportional to flow velocity. Figure 6.5 shows the principle of an electromagnetic flowmeter.

Even poorly conductive fluids can be measured using such a transducer and because it has no moving parts and has a smooth fluid tube electromagnetic flowmeters are highly reliable and can be used for slurries as well as liquids.

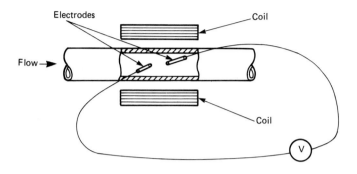

Figure 6.5 *Using the magnetic field produced by two coils to create an EMF according to fluid flow, in the electromagnetic flow transducer*

Vortex generating flow sensing

Any obstruction in the fluid pipe will make the fluid swirl in a vortex (Figure 6.6). The vortex is proportional to the volumetric flow rate of the fluid. Two methods of vortex generation are used: **forced oscillation** (Figure 6.6(a)), in which the fluid rotates or *precesses* along the axis of the fluid tube in a sort of spiral; **natural oscillation** (Figure 6.6(b)), in which a stable pattern (known as a *von Karmann vortex street pattern*) of alternate fluid rotations are *shed* from the obstruction.

Figure 6.6 *Principles of vortex generating flow transducers: (a) forced oscillation, in which the fluid precesses; (b) natural oscillation, in which the fluid rotations are shed from the obstruction*

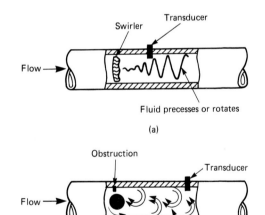

Forced oscillation vortex flowmeters generally use a piezoelectric transducer to sense the number of precessions passing a point. Natural oscillation vortex flowmeters typically use a strain gauge force sensor or an ultrasonic means to detect the periodic fluctuations of force which the vortex street pattern produces.

Ultrasonic flow sensing
Apart from being used as transduction elements of natural oscillation vortex generating flowmeters, ultrasonic transducers can be used to sense flow in their own right. Various techniques exist, for instance, measuring the time it takes an ultrasonic pulse to travel through the fluid. Doppler effect can also be used, the frequency of a received ultrasonic pulse varying with velocity of fluid flow.

Pressure
Transducers to measure pressure work on some mechanical principle, using an elastic section of material which reacts by movement when the applied pressure acts upon it.

Measurement of this movement (or measurement of the strain which the material experiences due to the pressure) provides an indication of the pressure. Typical sensing methods are illustrated in Figure 6.7.

Figure 6.7(a) shows a diaphragm mounted such that the pressure to be measured acts only on one side. The diaphragm can be corrugated, as shown, or flat. If two corrugated diaphragms are mounted back to back (Figure 6.7(b)), and the pressure to be sensed is piped into their centre, a more effective sensing element called a **capsule** is produced. Figure 6.7(c) shows a **bellows**. Finally, Figure 6.7(d) to (f) shows different types of Bourdon tubes, which move with applied pressure.

There are three ways in which pressure is measured: **absolute**, **differential**, or **gauge pressure**. Absolute pressure is the pressure of the fluid with respect to a vacuum. Differential pressure is the fluid pressure with respect to a

Plate 6.4
Ultrasonic flowmeter (Krohne Measurement & Control)

Plate 6.5
Ultrasonic flowmeter which can be clamped on to a metal pipe (Krohne Mesurement & Control)

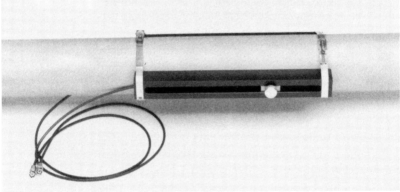

nonzero reference pressure. Gauge pressure is fluid pressure with respect to atmospheric pressure. As all the pressure sensing methods are *really* measuring differential pressure anyway, i.e., the difference in pressure between one side of the element and the other, it becomes a simple job to determine which type of pressure (absolute, differential or gauge) is indicated by the transducer. All that is

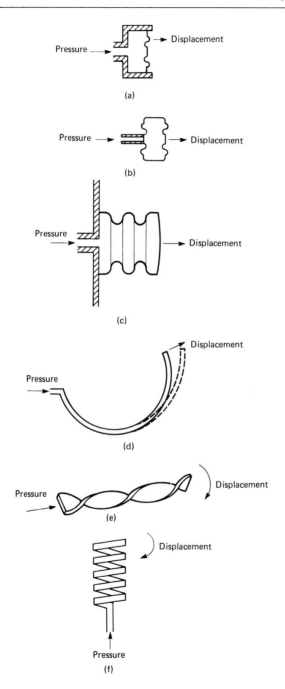

Figure 6.7
Pressure transduction methods: (a) a diaphragm; (b) a capsule; (c) a bellows; (d) a simple Bourdon tube; (e) a twisted Bourdon tube; (f) a helix Bourdon tube

required is to fix the pressure at one side of the sensing element with a vacuum, the pressure to be compared with, or atmospheric pressure. Figure 6.8 illustrates the procedure.

Figure 6.8(a) shows how an absolute pressure measurement can be made. Figure 6.8(b) shows differential pressure measurement, while Figure 6.8(c) shows gauge pressure measurement. The designer of the pressure transducer simply chooses the pressure indication required.

Strain gauge pressure transducers
By attaching a strain gauge of some description to the sensing element, a straightforward pressure transducer results. As the element is typically circular, a rosette type of strain gauge is usually required, although by no means essential.

Thin-film strain gauges allow a robust transducer to be constructed, while bonded semiconductor gauges or even integral diffused gauges allow rapid response and high accuracy.

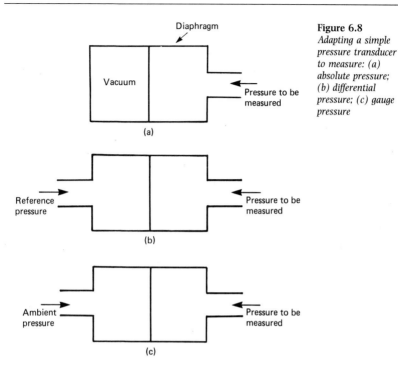

Figure 6.8
Adapting a simple pressure transducer to measure: (a) absolute pressure; (b) differential pressure; (c) gauge pressure

Capacitive pressure transducers

One of two main designs is used in capacitive pressure transducers: **single-stator**, in which the sensing element (usually a diaphragm) is made one electrode of the capacitor, and moves to and from a stationary electrode (Figure 6.9(a)); **dual-stator**, in which the diaphragm moves between two stationary electrodes (Figure 6.9(b)).

In varying gas pressure measurements it is possible to use a simple capacitive microphone as a pressure transducer. Frequency response is usually the deciding factor, though, although some types are available with response virtually down to 0 Hz, and as high as 200 kHz.

Reluctive pressure transducers

The linear variable differential transformer (LVDT) forms the basis of the most commonly used reluctive pressure transducer. In effect, the transducer is simply a mechani-

Figure 6.9 *(a)*
single stator; (b)
dual capacitive
pressure transducers

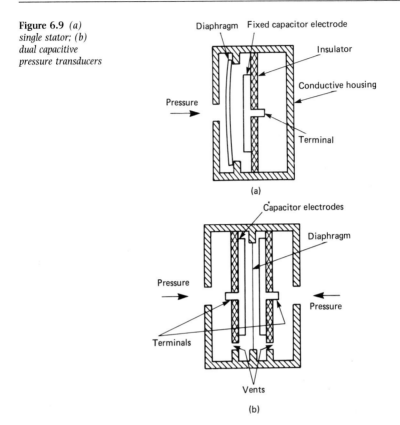

cal sensing element, whose movement is measured by the LVDT (Figure 6.10). Typically, capsules, bellows or Bourdon tubes are used as the sensing elements.

Of the other reluctive transduction methods, a twin-coil inductance bridge is sometimes used, as in Figure 6.11. A single diaphragm separates two coils, so that upon diaphragm movement the inductance of one coil reduces while that of the other increases.

Pressure switches

Often, it is not necessary to measure the pressure in question, and a simple switchover when a certain point of pressure is passed is all that is required.

Plate 6.7
*Transducers for use
with the flowmeter
of Plate 6.6*
(Detectronic)

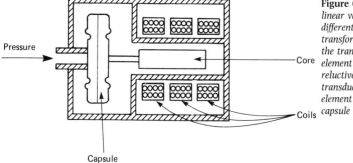

Pressure

Core

Coils

Capsule

Figure 6.10 *A
linear variable
differential
transformer used as
the transduction
element in a
reluctive pressure
transducer. Sensing
element is a simple
capsule*

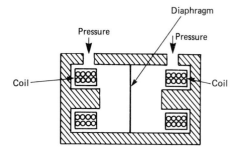

Diaphragm

Pressure

Pressure

Coil

Coil

Figure 6.11 *Twin
coil inductance
bridge pressure
transducer*

Plate 6.8
Transducers for use
with the flowmeter
of Plate 6.6
(Detectronic)

A typical application of such a device can be seen in all modern cars, where a pressure sensor is used to monitor oil pressure – if a leak occurs causing oil pressure to fall, the monitor switches on a dashboard light to warn the driver. In an example like this the predetermined switchover point is fixed at manufacture stage, but some transducers allow the switchover point to be adjusted by the user.

Humidity

A sensing element which has some measurable property varying with humidity must be used in a humidity transducer. Sensing elements can be as simple as a hygroscopic element (i.e., one which adsorbs water molecules) held taught by a spring arrangement. Any adsorption or desorption of moisture alters the length of the element and so any marked point on the arrangement moves in one direction or the other according to the level of adsorption – movement is then sensed by one of the common transduction principles. Although inorganic hygroscopic elements are available, it is possible to use certain animal membranes or even human hair.

Resistive hygroscopic elements employ a change in resistance to sense humidity change. They are, essentially, a wire element coated with an aqueous salt solution. The salt layer changes resistance with local humidity. Hygrometer elements can also be manufactured in which the surface itself changes resistance. One example is the *Pope element* comprising polystyrene treated with sulphuric acid.

Aluminium oxide hygrometric elements change capacitance as well as resistance with a corresponding change in humidity. Construction is of an aluminium substrate forming one electrode, with a layer of aluminium oxide, and a thin layer of gold (thin enough to be porous) to form

Plate 6.9
Piezoresistive element gauge pressure transducer
(RS Components)

the other electrode of the capacitive element. The aluminium oxide's structure is of thin pores of material which soak up water vapour, changing the dielectric constant of the capacitor and hence its capacitance.

If a quartz crystal is coated with a hygroscopic material coating its resonant frequency will change with humidity, because humidity affects the overall mass.

Level

Liquid level sensing in terms of switching is a relatively simple task, and is known as discrete level sensing (as opposed to continuous level sensing – see later). The most common example (non-electronic, of course), is the conventional ballcock lavatory cistern device, which is merely a stop valve whose water output is stopped when a floating ball rises high enough to close the valve. This is known as float level sensing. In electronic terms, transducers are typically every bit as simple: when a float rises above a certain point, switch contacts close. Often, the switching arrangement is nothing more than a reed switch mounted in the transducer body, and a permanent magnet mounted in the float. Many float level sensing transducers are available.

Continuous level sensing float transducers are available which use the position of the float to change a variable transduction element, such that output signal is indicative of the actual level rather than just switching at a desired point. Fuel sender unit of a car's petrol tank is a good example of such a transducer, where the float position adjusts a variable resistance.

Conductivity level sensing

If electrodes are inserted in the fluid whose level is to be measured, then a change in conductivity is indicative of liquid level. Current through the fluid must be sufficiently small to eliminate the possibility of electrolysis or explosion.

This method relies on the fluid being conductive, of course.

Capacitive level sensing

Similar to conductive level sensing, in that electrodes are inserted in to the fluid, but using the fluid as a dielectric between the two electrodes to form a capacitor (Figure 6.12). Changes in the fluid level mean that the dielectric constant changes, and so the capacitance of the transducer so formed changes. Unlike conductive level sensing transducers, capacitive level sensing transducers can only be used with nonconductive fluids.

If the container is metallic, it is possible to use it as one electrode of the transducer.

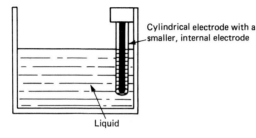

Cylindrical electrode with a smaller, internal electrode

Liquid

Figure 6.12
Capacitive liquid level sensing. The transducer capacitor dielectric is formed by a liquid layer and, above that, gas or air. As liquid level changes so does the ratio of liquid to air forming the dielectric, therefore the capacitance changes

Photoelectric level sensing

Generally, it is only possible to use photoelectric sensing methods in discrete level sensing transducers. Two main modes of operation are used and they are shown in Figure 6.13. The first uses physically separate photoelectric sources and detectors, such that the photoelectric beam between the two is cut off when the liquid whose level is being measured comes between them (Figure 6.13(a)). In practice the beam is not totally cut-off, but attenuated –

Plate 6.10
Photoelectric level sensing transducers with integral LED source and photosensor, operating on the total internal reflection principle. When liquid level is above the transducers, light is not internally reflected (RS Components)

Figure 6.13 *Photo-electrical liquid level sensing principles: (a) as liquid level rises above the level of the light beam between source and detector, the beam is cut off and no longer detected; (b) as liquid level rises above the commonly housed source and detector, the refractive index between the prism and its surroundings alters sufficiently so that light is not reflected back to the detector*

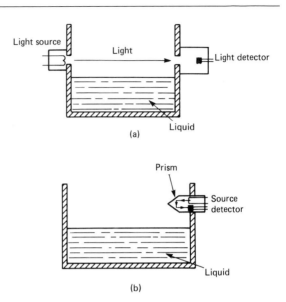

following interface circuits must be used to determine the switch point.

The second main mode uses a single housing for source and detector, together with a prism arrangement (Figure 6.13(b)). Light from the source is reflected internally back to the detector when the housing is in gas. When the liquid level is sufficient to cover the housing, the index of refraction between the prism and its surroundings changes so light is not reflected back to the detector. Technically, of course, light is not the only form of electromagnetic radiation which can be used in photoelectric level sensing – ultraviolet and infrared radiation can be used, too.

Ultrasonic level sensing
Ultrasonic level sensing methods can be used to give continuous level as well as discrete level monitoring. There are three main operating modes. First is similar in principle to the first photoelectric level sensing method, in that an ultrasonic emitter and detector are mounted so that a direct wavepath exists between the two in gas (Figure

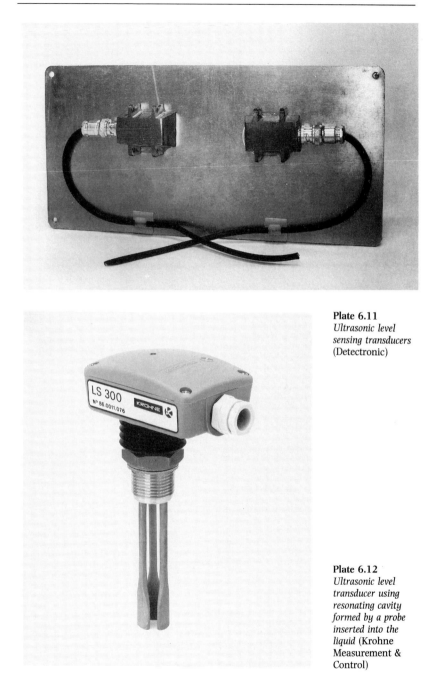

Plate 6.11
*Ultrasonic level
sensing transducers*
(Detectronic)

Plate 6.12
*Ultrasonic level
transducer using
resonating cavity
formed by a probe
inserted into the
liquid* (Krohne
Measurement &
Control)

Figure 6.14
Principal methods of ultrasonic liquid level sensing: (a) liquid cuts off the ultrasonic path when high enough; (b) the ultrasonic beam is only reflected back from emitter to detector when liquid level is at a particular point; (c) cavity resonance method – different liquid levels produce different resonant frequencies

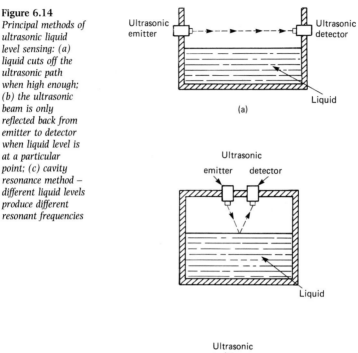

(a)

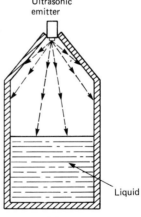

(b)

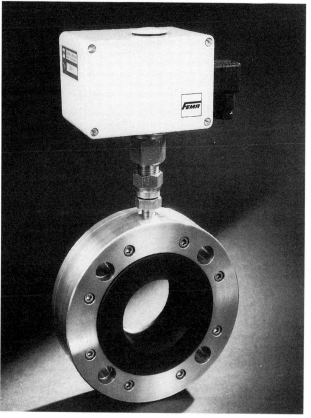

Plate 6.13
*Hydrostatic level
transducer used to
measure the level of
slurries, comprising
diaphragm seal and
pressure transducer*
(Texcel)

6.14(a)). When liquid level covers the ultrasonic devices, the ultrasonic wave is sufficiently attenuated to pass a predetermined switch point. Like the first photoelectric sensing method, too, this allows only discrete level sensing.

By reflecting the ultrasonic wave off the surface of the liquid (Figure 6.14(b)), the second ultrasonic level sensing method can be used for continuous level sensing. Pulses are transmitted at the liquid level, where they bounce off and are reflected back to the detector: the time measured between emission and detection is indicative of the distance of the liquid level from the transducer. Calculation of the liquid level must take into account the velocity of sound in the medium between the transducer and the surface of the liquid. Transducers may be formed by separate emitters and detectors (as shown) or a single transducer doubling as emitter and detector may be used.

A third ultrasonic level sensing method is shown in Figure 6.14(c), where a single emitter is used to direct ultrasonic waves into the cavity above the liquid, so that the waves occur at the cavity's resonant frequency, or at a harmonic of that frequency. At a different liquid level the resonant frequency is correspondingly different, so it is a matter of measuring the new oscillation frequency to determine the liquid level.

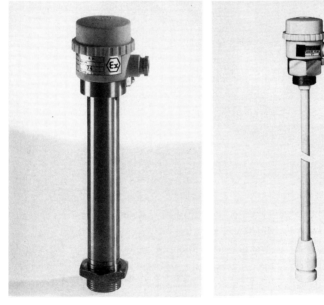

Plate 6.14 *Hydrostatic pressure probe, in explosion-proof housing for continuous level monitoring* (Texcel)

Plate 6.15 *Capacitance rod probe for continuous level monitoring* (Texcel)

7

Acoustic and optical measurands

Acoustics

The basic measurand in acoustics is sound, although there are associated measurands derived from this – the main one being sound pressure level. Two other measurands to be considered are those of sound travelling in water and ultrasonic sound.

Sound

In air, sound is simply a collection of pressure waves, moving at the velocity of approximately 322ms^{-1} at sea level. These pressure waves of air act at a point as oscillations of air pressure. Strictly, these oscillations must occur within the human frequency range of hearing (approximately 30 Hz to 15 000 kHz) to be heard and therefore be classed as the phenomenon of sound.

Air is an elastic medium. Sound can travel through any other elastic medium, too, although it might not travel as oscillations of pressure, rather oscillations of particle displacement, stress, or density. Further, it travels at different velocities through different media.

It follows that, in air at least, measurement of sound is accomplished using pressure sensing transducers with a frequency response equal to that of human hearing. Such transducers are usually called **microphones**.

Sound pressure level

Sound pressure level is normally expressed in decibels, as:

$$\text{SPL} = 20 \log_{10} \frac{\rho(\text{rms})}{\rho_{\text{ref}}(\text{rms})}$$

Where $\rho(\text{rms})$ is the effective sound pressure, and $p_{\text{ref}}(\text{rms})$

is a reference sound pressure. The reference pressure should be stated.

Sound level

Sound pressure level measurements involve direct readings of the pressures of all frequencies of sounds such that the noted pressure levels are exactly those which occur. They are quantitive measurements, objectively obtained. *Sound level* measurements, on the other hand, are **weighted**: the microphone signal passes through a weighting network which emphasizes signal components of certain frequences while de-emphasizing those of other frequencies. Purpose behind this is to correlate sound measurements more closely to the *subjective* sensation of sound which humans detect – sound pressure level measurement does not (and cannot) take human sound sensation into account. National standards define characteristics of weighting networks and designate a reference letter (for example A, B, C) to be used with a particular characteristic.

Sound level meters and sound pressure level meters are available as self-contained hand-held units, complete with weighting networks, preamplifiers, and displays, ready for use in sound measurement applications.

Microphones

Sound pressure varies over an extremely wide, but nevertheless low, range. Frequency response is also wide. Transducer sensing elements must therefore be reasonably stiff and of low mass, with small deflection. Generally a flat diaphragm pressure sensing element is used in a gauge pressure configuration (see Chapter 6).

The most common types of microphones used in sound measurement systems are crystal (working on the piezoelectric principle) and condenser (working on the capacitive principle) microphones. Other types of microphones exist, however.

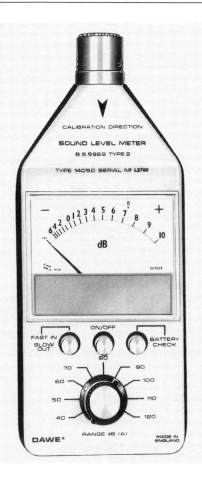

Plate 7.1 *Sound level meter, with integral microphone, weighting networks, display etc.* (RS Components)

Piezoelectric microphones

The diaphragm of the microphone is mechanically coupled to, and acting directly on, a small piece of ceramic or quartz crystal (Figure 7.1), so that the output signal obtained from the crystal by piezoelectric effect is proportional to the sound impinging on the diaphragm.

Often, microphones are equipped with an integral preamplifier which is used as an impedance converter, to give a lower output impedance (anywhere from around $100\,\Omega$ to $10\,000\,\Omega$) than the crystal unit provides (in the order of megohms). This helps the engineer to keep noise pickup to an acceptable level in long transmission-path instrumentation systems. Power to the amplifier can be provided along the connecting cable from the rest of the system, or may be from an integral cell within the microphone body.

Figure 7.1
Piezoelectric microphone. The diaphragm is mechanically linked to the crystal element

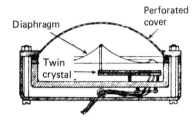

Condenser microphones

A basic condenser microphone make-up is shown in Figure 7.2, where a diaphragm, electrically connected to the body of the microphone, acts as the sound pressure sensing element and also forms one electrode of the capacitor. A backplate forms the other electrode.

Figure 7.2 *Internal view of a basic condenser microphone*

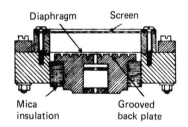

If the microphone is of this basic form a charge must be maintained on the two electrodes, and this is typically provided by a DC voltage of around 200 V from a separate power supply unit. Some forms of condenser microphone sidestep this problem by introducing a polarized dielectric between the electrodes. The polarized dielectric is known as an **electret** and so this form of condenser microphone is commonly called an electret microphone. Electret microphones are usually supplied with an integral field effect transistor preamplifier, powered by a small cell, which acts as an impedance converter (see previous section on piezoelectric microphones).

Hydrophonics

Sound travelling in water is measured using a version of microphone known as the **hydrophone**. Piezoelectric construction is the most common form, similar to conventional piezoelectric microphones.

Sound sources can be located using two or more hydrophones in a process known as **ranging**. Often, two hydrophones are used, one as sound emitter and one as sound detector in echo-ranging equipment: where pulses of sound are emitted and detected after reflection (i.e., echo) from an underwater body. The time taken between emission and detection is indicative of the distance from the body.

Ultrasonics

Ultrasonic waves are similar to sound waves in that they are, in air, moving pressure waves. They differ from sound only in frequency: ranging from about 20 kHz upwards. No particular upper limit is imposed on ultrasonic waves – as long as the wave propagates in a medium then it can be used.

Ultrasonic transducers are available allowing transmission or detection of ultrasonic waves, at a variety of frequencies. They are, typically, piezoelectric microphones specifically manufactured to have a resonant frequency within the ultrasound range. Connecting the emitting transducer to an oscillator of the same resonant frequency as the emitter will cause the emission of ultrasound waves. Emitters and detectors matched in resonant frequency must be used.

Similar ranging and echo ranging techniques to those of hydrophonics may be used with ultrasound transducers to allow many investigations of industrial or medical forms.

Plate 7.2
Ultrasonic (40 kHz)
transducer (RS
Components)

The advantage of ultrasound, however, is that it is far more directional than sound, giving a quality which is much more light-like in reflectional and interactional characteristics. The best known example of ultrasound capabilities is in medical ultrasound scanning: where babies can be observed in uteri by external examination.

Optics
Light is an electromagnetic radiation.

Like sound, light can only be strictly defined in terms of human subjectivity. That is, *visible* light is radiation in the wavelength range which the human eye can detect – about 380 to 780 nm. However, wavelengths above and below that of light display light-like characteristics which may be used effectively in many applications.

Wavelengths of radiation between about 10 and 380 nm are strictly called **ultraviolet** (UV) **radiation**, although they are commonly called ultraviolet *light*. Similarly, radiation wavelengths of between about 780 to 3000 nm and beyond are called **infrared** (IR) **radiation**, although they are often, incorrectly, called infrared light. By far the most important measurand to be considered in optical measurements is that of intensity, although colour (for light) can be detected.

Intensity

Transducers used to measure optical quantities are referred to as sensors or, sometimes, detectors. They should also, correctly, be specifically named as either IV, IR or light sensors or detectors. Sometimes the terms photosensors, photodetectors, or optoelectronic devices may be used, but these are generic terms for all optical transducers and give no reference to which wavelength is actually measured.

Optical transducers can be classified into two main categories: photon detectors (from where the term 'photosensors' originates); and thermal detectors.

Photon detectors

Main photon detection sensing methods are shown in Figure 7.3.

- **Photoconductive** sensors (Figure 7.3(a)) consist of a semiconductor material which has a resistance proportional to illumination. Incident photon energy is absorbed by the semiconductor producing a change in the number of charge carriers, thus changing the resistance accordingly. Photodiodes (Figure 7.3(b)) and phototransistors (Figure 7.3(c)) are examples of **photoconductive junction** sensors, in which the photoconductive semiconductor is used in the junction of a diode or transistor. The PIN diode (the name arises from the fact that the diode is a junction between p-type and n-type semiconductors, with a separating layer or pure, or intrinsic, semiconductor) of Figure 7.3(d) is a variation.
- **Photoemissive** sensors emit electrons when photons are

incident upon either a cathode (in the case of a photomultiplier – Figure 7.3(e)) or a p-n junction (in the case of an avalanche photodiode – Figure 7.3(f)).

Photomultipliers are usually housed in a vacuum tube arrangement and have sets of dynodes maintained at increasingly higher positive potentials. As photons enter the photomultiplier and strike the cathode, electrons are released by photoemission, and these are attracted to the positive potentials of the dynodes. Upon striking the first dynode, each electron causes two or more electrons to be released by the process of secondary emission. Amplification of the electrons thus takes place at each dynode.

Avalanche photodiodes use a similar effect with the charge carriers of a p-n junction which is maintained at a high reverse bias. Incident photons split hole/electron charge carriers so that the electrons travel toward the n-type layer of the junction and holes travel toward the

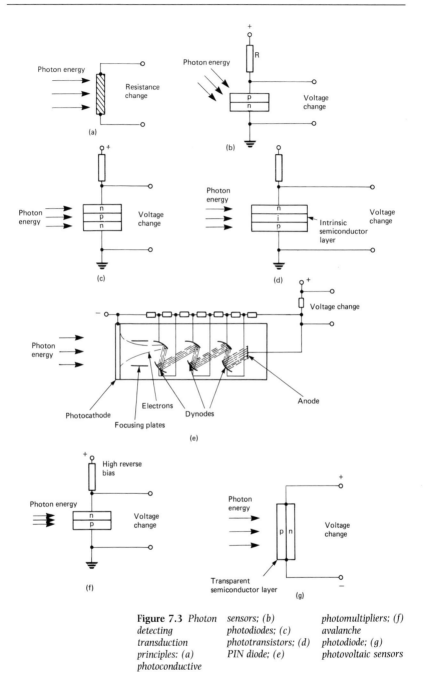

Figure 7.3 *Photon detecting transduction principles: (a) photoconductive sensors; (b) photodiodes; (c) phototransistors; (d) PIN diode; (e) photomultipliers; (f) avalanche photodiode; (g) photovoltaic sensors*

p-type layer. The high reverse bias ensures that each hole or electron is travelling with sufficient energy to split further hole/electron pairs in an avalanche effect.

- **Photovoltaic** sensors (Figure 7.3(g)) are similar to basic photodiodes in that a p-n junction is used to detect incident photons. However, the junction is unbiased and when connected to a load circuit causes current to flow according to the amount of incident light. Some devices may be used in either the photodiode or photovoltaic modes.

Thermal detectors

Three main types of thermal detectors exist: bolometric, pyroelectric and thermoelectric transducers. They all function in the same way as devices used for radiation pyrometry (see Chapter 4) and so will not be covered in detail here.

8
Chemical measurands

The subject of the measurement of chemical measurands, more so than any other category of measurands, is not one which can be covered in detail by a book of this sort. There are so many possible measurands that engineers involved in chemical industry are usually forced into designing and making their own transducers. Nevertheless, there is a number of measurands whose measurements may be listed because they allow a central core of transducers to be formed. These measurands are: acidity; conductivity; oxidation-reduction potential.

Chemical analysis of materials is quite another problem. It would not be possible in a single book (let alone this chapter) to gather together all possible transducers capable of measuring levels of substances in a sample – often no particular transducer is used: the analysis being more of a complete process than a single measurement. Sometimes, however, certain types of transducers are used so regularly that they become commonplace. Good examples of such transducers are those used in the smoke alarms which are available, for home use, for just a few pounds.

Most of the transducers covered here allow a sample of the substance under observation to be taken into a probe of some sort. Inside the probe, conditions for a chemical reaction are made available and the electrical characteristics which are then measured provide the required information about the original sample. Transducers which work in this way are known as **electrometric**.

Acidity
The acidity, or alkalinity, of a solution is denoted by its

hydrogen potential (abbreviated to pH) value; where the pH value is given by the formula:

$$pH = -\log_{10}[H^+]$$

Where H^+ is the hydrogen ion concentration in grams per litre.

The pH value is expressed by numbers in the scale from 0 to 14. Pure water has a pH of 7, i.e., it is neutral, neither acid nor alkaline. A pH of 0 indicates a strong acid solution, while a pH of 14 indicates a strong alkaline solution. By retrospect we can see, therefore, that an acid has a greater concentration of active hydrogen ions than does an alkali.

Electrometric measurement of the solution whose acidity is being measured usually means inserting two special electrodes into the solution. Measurement of the potential developed across them gives an indication of pH value of the solution. One of these electrodes is known as the pH electrode (Figure 8.1(a)), the other is a reference electrode (Figure 8.1(b)).

Figure 8.1 *pH measuring transducer: (a) a pH electrode, used in coordination with; (b) a reference electrode*

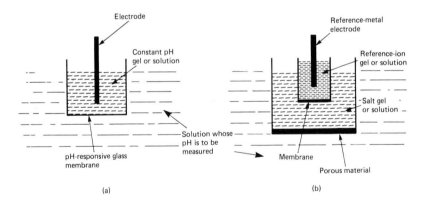

(a)   (b)

Modern pH electrodes are in the form of a single probe (Figure 8.2), housing both electrodes, known as a **combination electrode**. A special glass membrane sensitive to pH value covers the probe tip, so that there is no need to draw a sample of solution into the probe.

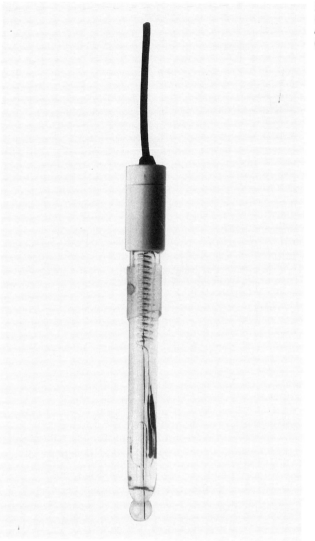

Plate 8.1
*Combination
electrode pH probe*
(RS Components)

Hand-held voltmeters are available specifically for pH value measurement and these display the pH value directly, although it is perfectly possible to use any voltmeter with a high enough input impedance (at least 100 MΩ) to make the measurement. Sensitivities of pH

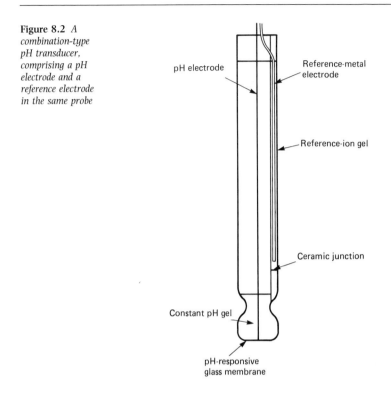

pH electrode

Reference-metal electrode

Reference-ion gel

Ceramic junction

Constant pH gel

pH-responsive glass membrane

electrodes are 59.1 mV for each unit of pH value, at 25 ° C. Care must be taken to compensate for temperature variations from this, as the thermal sensitivity of the transducer is about 0.2 mV for a unit of pH value for each degree Celsius. Modern pH meters typically use a temperature sensing transducer in the probe, or a variable control on the meter itself, to provide temperature compensation.

Oxidation-reduction potential
A similar arrangement to the pH electrode can be used to measure oxidation-reduction potential (redox). A positive redox value indicates the solution contains an oxidizing agent, a negative redox value indicates presence of a reducing agent.

Plate 8.2 *pH meter, comprising integral voltmeter with variable temperature compensation control* (RS Components)

Specific-ion transducers

Oxidation-reduction potential electrodes and pH electrodes are examples of transducers which can detect and allow measurements of specific ions in solution. Transducers capable of the same job with other ions are available. All use the same basic make-up of a sensing electrode and a reference electrode immersed in the solution, either as two separate probes or one combination-type probe. It is the nature of the two electrodes, however, which determines the ion to be specifically measured.

Conductivity

Measurements of conductivity of solutions can help in the determinations of concentrations. Basic principle is that of electrolysis, i.e., two electrodes are inserted into the solution and a voltage applied across them. The voltage causes the compounds in solution to separate into ions, which migrate towards the electrodes, effectively forming a current. Measurement of the current flowing through the circuit allows a calculation of the solution's conductance to be made. Conductance, G, is the reciprocal of resistance, R, where:

$$G = \frac{1}{R} = \frac{I}{V}$$

Where I is the current through the circuit and V is the voltage across the electrodes. Conductance has the units of siemens (S).

If the two electrodes have an effective surface area, A, and are separated by the distance, D, then the solution's *conductivity* is given by:

$$\gamma = \frac{GD}{A}$$

Where the two electrodes are mounted in a probe (the usual method), the ratio D/A is, of course, a constant – known as the **electrode constant**, or the **cell constant**.

The type of voltage used in conductivity measurements is of importance – a pure DC voltage would cause sufficient electrolytic action to occur at the electrodes to lower the current. Current measurement at this time would give a false conductivity measurement. An AC squarewave voltage is normally used.

Electrometric gas analysis

Electrometric gas analysis transducers are regularly used to determine the amount of a specific gas in a gas mixture or solution. One example is the zirconium dioxide exhaust gas transducer, used in automotive engine control systems, particularly in the USA. Figure 8.3 illustrates the principle behind its operation, where a tube of zirconium dioxide is precoated, inside and out, with porous platinum to form electrical contacts. A reference gas with a known amount of oxygen is put in the tube and each end of the tube is sealed with an insulating stopper. Alternatively, the reference gas chamber may be vented to allow ambient air infiltration.

A heater around the tube heats it to a temperature over $400\,^{\circ}C$; when the oxygen ions in the zirconium dioxide become mobile and the tube body effectively becomes an electrolytic conductor. The side of the tube in contact with the gas of lower oxygen content becomes negative with respect to the other side: the potential so formed is proportional to the relative oxygen contents of the two gases.

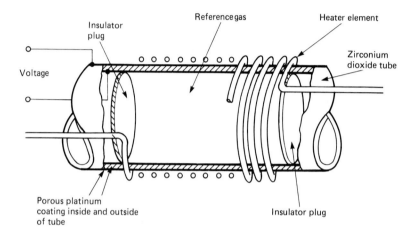

In the case of an engine exhaust gas transducer, the transducer is inserted into the exhaust system of the engine so that the outside of the zirconium dioxide tube is in contact with engine exhaust gas. Theoretically, the exhaust gas of an engine should have little, or no oxygen content; all oxygen being used up in the combustion process. Presence of oxygen, indicated by a change in voltage, would therefore indicate incomplete combustion.

Figure 8.3 *Zirconium dioxide exhaust gas transducer. An example of electrometric gas analysing transduction*

Resistive gas analysis

A resistive oxygen transducer such as the titanium oxide transducer can also be used as the basis of an exhaust gas analyser. Titanium oxide is a substance whose resistance changes according to the number of oxygen molecules absorbed in its surface.

The transducer is formed by a platinum wire or thin-film

Plate 8.3 *Resistive gas analysis transducer for detecting presence of propane and methane and other gases* (RS Components)

resistor, whose surface is coated with titanium oxide. According to the oxygen content of the ambient gas, so the titanium oxide layer changes resistance and so the overall resistance of the transducer changes.

Other substances can be used to create transducers capable of detecting other specific gases. Available transducers include those capable of detecting propane and methane. Often, resistive transducers have two elements: one coated; the other uncoated for use as a temperature compensating element when measurements are taken with the transducer in a bridge circuit.

9
Interfacing

Transducer interfacing is a knotty problem, not the least because there are so many types of transducer. Fortunately the circuits used are often similar for a number of types of transducer and so it is easy to generalize about them. At the same time, the engineer must still be aware of transducer differences so that the circuits may be adapted correctly. Problems which affect measurement systems differently, such as noise and interference are also discussed.

Many of the principles involved in interfacing transducers can be understood with reference to resistive transducers. Resistive transducers are those whose resistance changes for some change in measurand. Usually, the interfacing circuits are used to alter the change in resistance to one of voltage. This voltage then forms the input to the remainder of the system.

There is a number of ways in which the resistance change can be converted to a voltage change. The simplest is the voltage divider (Figure 9.1(a)), where the transducer resistance, R_t, is joined in a series circuit with another resistance, R_1, and an excitation voltage, V_{exc}. Output voltage, V_{out}, varies with transducer resistance according to the voltage divider formula:

$$V_{out} = V_{exc} \frac{R_1}{R_1 + R_t}$$

As the varying transducer resistance implies a varying load on the excitation supply, it is normally best to use a constant current power supply for excitation purposes.

Indeed, if a constant current supply is used, there is no need to use a series resistance at all – the voltage generated across the transducer resistance is directly measurable (Figure 9.1(b)).

Figure 9.1
*Interfacing resistive
transducers by
changing resistance
to voltage, using: (a)
the voltage divider;
(b) a constant
current power
supply; (c)
unbalanced
Wheatstone bridge*

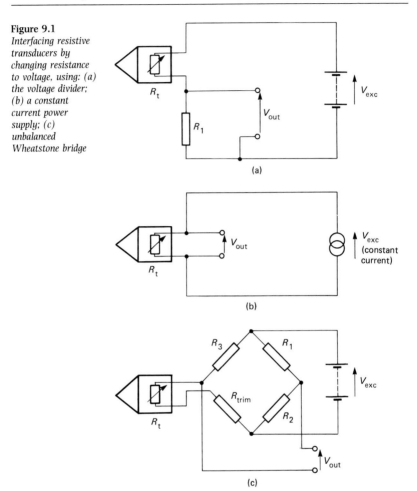

Most usual interfacing method for resistive transducers, however, is the unbalanced bridge arrangement of Figure 9.1(c), where the transducer resistance forms one arm of a Wheatstone bridge. If the transducer has more than one transduction element it is perfectly feasible to connect all into the bridge, too. Usually, a trimming resistor, R_{trim}, is included in series with the transducer so that the bridge may be balanced at some point (say, the lowest resistance point) in the range measurand being measured.

Output voltage is given by:

$$V_{out} = V_{exc} \frac{R_3}{R_3 + R_t + R_{trim}} - \frac{R_1}{R_1 + R_2}$$

In practice, this arrangement is often called a **strain gauge bridge** because it is used most often with strain gauge transducers. Typically, one, two or even four transduction elements in the same strain gauge transducer are connected into the bridge. Figure 9.2 (a) shows four possible strain gauge elements in a bridge layout, with upwards arrows indicating increasing resistance and downward arrows indicating decreasing resistance. Figure 9.2(b) shows the elements in a possible mechanical form, where the sensing link moves either to the left or to the right depending on how strain is applied to the device. The indicated directions of resistance changes in Figure 9.2(a) will occur in the device of Figure 9.2(b) if the sensing element is moved to the left upon application of strain.

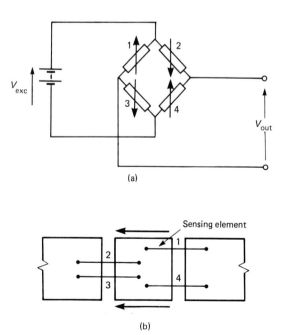

(a)

(b)

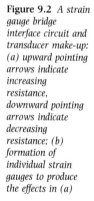

Figure 9.2 *A strain gauge bridge interface circuit and transducer make-up: (a) upward pointing arrows indicate increasing resistance, downward pointing arrows indicate decreasing resistance; (b) formation of individual strain gauges to produce the effects in (a)*

Strain gauge temperature compensation

A problem arises when using a strain gauge to measure strain on a surface when different temperatures may be encountered – differential expansion. If, say, the temperature rises, the surface being measured would expand a certain amount, which may not be the same amount by which the gauge element would naturally expand. But, as the element is fixed to the surface, the gauge is forced to expand by the same amount the surface expands, causing strains in the gauge and a consequent resistance change.

In an attempt to avoid this error, some gauges are manufactured so that resistance change due to differential expansion balances out the basic resistance change due to temperature (they are, in fact, of opposite effect). This can only be done, however, if the gauge is used only with the material for which it is designed. Manufacturers usually offer this compensation in gauges designed for use on the surface of one of three materials: stainless steel; mild steel; aluminium.

Bridge circuit temperature compensation

The previous temperature error and associated compensation is specific to the strain gauge. One problem with temperature when using *any* transducer in a basic bridge circuit is encountered if long connecting leads between transducer and bridge are used. Resistance of any material – including that of the connecting leads – is temperature dependent. A change in ambient temperature can therefore introduce a change in bridge output voltage.

Fortunately, compensation of this kind of temperature error is quite easy. Figure 9.3(a) shows one method, where a **three-wire** transducer is used. All three connecting leads are of the same length and so have the same resistance. Thus any change in the resistance of the transducer arm of the bridge is compensated for by an identical resistance change in the R_2 bridge arm.

Another temperature compensation method is shown in Figure 9.3(b), where a compensating arm of the bridge

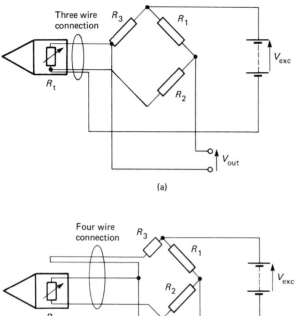

Figure 9.3 *Two methods of compensating for temperature variations with a strain gauge bridge circuit: (a) using a three-wire transducer; (b) using a four-wire connection. In both methods the connecting wires should be twisted lightly*

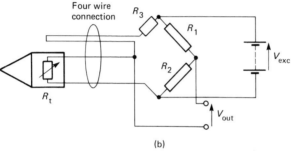

is wired in a loop, parallel to the transducer connection. Resistance change due to temperature variations thus affect both arms of the bridge identically.

In both these temperature compensation methods, connecting leads should be twisted lightly to ensure temperature variations along the length of the connections affect all wires equally.

Installation of strain gauges

Strain gauges differ from most other types of transducer in that their installation onto the surface whose strain is to be measured is usually permanent. Typically an epoxy adhesive is used to 'cement' the element or device into place, and so care must be taken to ensure correct fixing – once in position it cannot be removed and refixed. The following steps are general steps which should be taken when installing a strain gauge:

● Immediately before installation, clean the surface onto which the gauge is to be mounted and roughen the surface slightly.

● Select the correct adhesive for the application; check that the adhesive is suitable for use with the sensor and surface, bearing in mind environment, humidity, temperature etc.

● After correct application of adhesive (following the manufacturer's instructions) clamp the gauge to the surface, using a metal plate, with a strip of nonadhesive plastic or similar material between plate and gauge; then apply even pressure over the installation and leave for the manufacturer's recommended curing time for the adhesive.

● After adhesive has cured, remove the clamping devices and apply a suitable moisture-proofing compound over the gauge.

Connections to the individual elements in a strain gauge installation are best made through a suitable connecting block, bolted or cemented close to the gauge installation on the measured surface. Moisture-proofing compound should also be applied to the connecting block, in such an instance.

Noise

Apart from the problems of temperature mentioned previously, long connecting leads can exacerbate other problems, too. Most notable of these are the effects of noise which, indeed, can affect any transduction system performance.

In any system where a sensitive transducer is supplying a weak signal to the remainder of the system, any small ingress of noise will be amplified to the extent where it is potentially impossible to make a correct measurement. Further, noise problems can occur in a number of different ways.

In some cases the effects of noise encountered by a system can be reduced to an acceptable level by filtering, but in many respects this is the electronic equivalent of shutting the barn door after the horse has bolted. To be sure of good system performance, it is always wise to assume that noise of all types will be encountered, and to design the system so as to reduce, if not eliminate, the noise levels to the extent where they are negligible in the first place.

Interference

There are two main types of noise which can affect a system. Noise which is *picked up* primarily by the connecting leads between transducer and interface (but occurs also elsewhere in the system) is normally called **interference** and is, in fact, manmade noise. A good example of interference is the low frequency *hum* generated by an amplifier and loudspeaker arrangement such as a home hi-fi system. The hum, at a frequency of 50 Hz or 100 Hz, is generated initially within the amplifier because of the close proximity of the amplifier's power supply. The power supply is mains powered and so low frequency noise at 50 Hz and/or 100 Hz (if full-wave rectification takes place in the power supply) is picked up by the amplifier. The problem arises in the first place because the low magnitude signals which the record player cartridge generates must be amplified to produce sound from the loudspeaker (or, in the case of a measurement system, low magnitude signals which the transducer generates). Pickup of some sort occurs between the transducer and amplifier and the interference is amplified along with the signal.

Two main types of interference are common (although others exist), and are illustrated in Figures 9.4 and 9.5. **Capacitively coupled** interference, more commonly called **capacitive** interference occurs when two systems are separated. Signals are present within each system (Figure 9.4(a)), and ideally no interaction occurs between the two. However, as there are metal parts (wires, component

Figure 9.4
Illustrating capacitively coupled interference

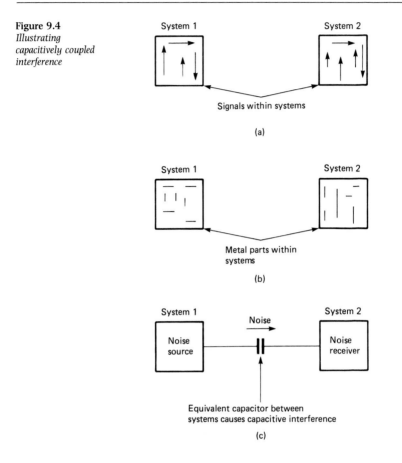

System 1   System 2

Signals within systems

(a)

System 1   System 2

Metal parts within systems

(b)

System 1   System 2

Noise

Noise source   Noise receiver

Equivalent capacitor between systems causes capacitive interference

(c)

bodies, cases etc.) in each system and as the metal parts in each system are connected by air (Figure 9.4(b)) the effect is that of a capacitor. An equivalent circuit of this arrangement (Figure 9.4(c)) shows that the capacitor acts like a coupling capacitor, and signals from one system can pass to the other as interference.

Figure 9.5 shows a similar arrangement in which two systems each with inductive components can act as if coupled by a transformer. This is known as **magnetically coupled** or **inductive** interference.

Within each type of interference there are a number of subcategories, usually named after the effects they produce. One important subcategory of capacitive interference is **capacitive hum** – as the previous example shows – which normally occurs because a power supply is in close proximity to, and is capacitively coupled with an amplifying circuit. Another subcategory is **capacitive crosstalk**, in which signals from one part of the *same* system are capacitively coupled to another part so causing interference. **Inductive hum** and **inductive crosstalk** are the magnetically coupled equivalents. Any or all of these interference types can be generated by a potential noise source such as a closely situated power supply, an electric motor, a badly suppressed engine, etc.

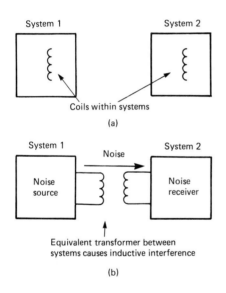

Figure 9.5
Illustrating inductively coupled interference

One of the first methods of attempting to eliminate interference is to connect transducer and interface circuits with screened or coaxial cable; earthing the screening braid in the belief that a shield is set up between a potential noise source and the system. However, this belief is not always justified and using earthed coaxial connections may make interference worse, not better. The reason for this is that a screen serves basically an electrostatic function, whose equivalent circuit is shown in Figure 9.6. Now, instead of only one capacitor coupling a potential noise source to the interface circuits (a potential noise receiver), two capacitors exist. If the electrostatic shield has zero resistance to earth then the equivalent circuit is one which directly earths the capacitively coupled noise signal – so no interference occurs. If, however, the shield has a finite resistance to earth (which is always the case

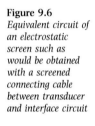

Figure 9.6
Equivalent circuit of an electrostatic screen such as would be obtained with a screened connecting cable between transducer and interface circuit

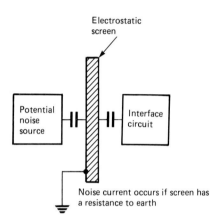

Electrostatic screen

Potential noise source

Interface circuit

Noise current occurs if screen has a resistance to earth

when a long connecting lead is used between a transducer and the rest of the system – purely because the cable has resistance) then the noise source will generate a noise voltage between the screen and earth. Interference still occurs and may be even worse than before. With short lengths of screened coaxial connecting cable this is not usually a problem, however.

Interference may also be caused by sloppy use of screened coaxial connecting cable, on the other hand, if the use of the term 'earth' is not fully understood. Figure 9.7(a) shows a transducer connected to an amplifier by coaxial cable, in which the coaxial cable's screen is earthed at the source *and* earthed at the receiver. Earthing, however, is not a guarantee that voltages at two different earth points will be identical. If even just a small difference in potential between the two earth points exists, then a

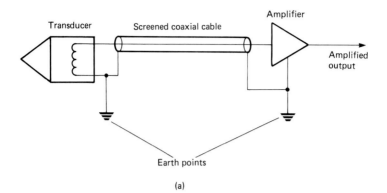

(a)

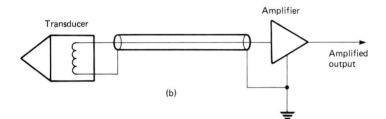

(b)

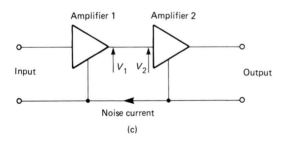

Noise current

(c)

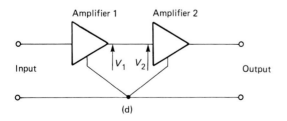

(d)

Figure 9.7
Earthing procedures:
(a) earthing
screened cable at
both ends could
create noise current;
(b) correct
connection of an
earth at only one
end of screened
connecting cable; (c)
noise current when
two sensitive
amplifiers have
separate earth
points; (d) correct
earthing
arrangement with
only one earth point

current will flow along the screen causing interference. This situation is known as an **earth loop**. In short, when screened coaxial cable is used to connect a signal source to a receiver, the screen must be connected at only one end (Figure 9.7(b)).

Earth loops can occur inside a single circuit, too, If, say, two amplifiers are in series and in close proximity, as shown in Figure 9.7(c), but each is earthed to its own earth point then the earths may have different potentials: so the output voltage of the first amplifier, V_1, is not the same as the input voltage of the second amplifier, V_2, and so noise current flows – even if the two earth points are formed by the same conductor which may be a single piece of printed circuit track. Indeed, the only safe way round this for sensitive circuits is to provide a common earthing point for all parts of the circuit, shown in Figure 9.7(d).

Figure 9.8
Balanced connection between transducer and amplifier, using a twisted wire pair to eliminate the effects of interference

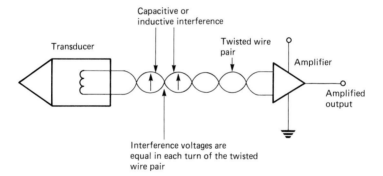

Another method of connecting a signal source to a receiver is that of twisted wire pairs in what is known as a **balanced** connection, illustrated in Figure 9.8. Interference affects each wire in the twisted pair, and because the wires *are* twisted the interferences are opposite at any one point: the overall effect being to cancel each other out.

Guard rings
Capacitive and inductive crosstalk, as well as simple resistive crosstalk, may occur if a high input impedance

amplifier is used in the interface. Such an amplifier is necessary to interface with a transducer having a similarly high output impedance. The amplifier's high input impedance means that any stray capacitances, stray inductances, or leakage resistances at its input will couple unwanted signals to the amplifier (Figure 9.9(a)). Modern operational amplifiers are particularly prone to this if used in such a configuration – input bias currents of only a few picoamps will cause significant interference. One solution is the **guard ring** approach (Figure 9.9(b)), in which the amplifier's high impedance input is enclosed with a low impedance guard, maintained at the same potential as the input. Typically, the high impedance amplifier is configured as a non-inverting amplifier (often called a **buffer**), such that its output signal is exactly the same as its input signal, but the amplifier's output impedance is much lower than that of its input. The guard ring is linked directly to the amplifier's output, so that it forms a low

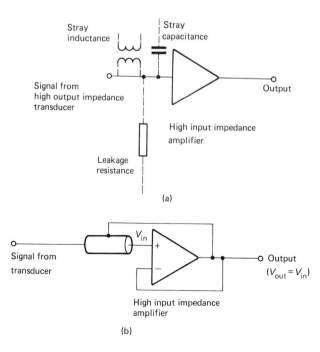

Figure 9.9 *Using a guard ring to reduce interference: (a) stray capacitances, inductances and leakage resistances cause interference at the input to a high input impedance amplifier; (b) usng a low impedance guard ring to reduce interference*

resistance path to signals from any stray capacitances, inductances, or leakage resistances. On printed circuit board the guard ring is formed by a large section of copper track, totally surrounding the amplifier, with single-strand wire links forming required power and other connections.

Random noise

The other form of noise is known as **random noise**, sometimes called **fundamental noise**. It is noise present in the system itself, caused by the basic physical properties of the components in the system. Where interference always has some distinguishable pattern or form, random noise has not, so potentially it is more difficult to remove. On the other hand, careful system design may be used in an attempt to keep the noise to a low, acceptable, level.

It is usually convenient to think of the noise generated by a system in the form of a ratio between the wanted signal and the unwanted noise. This is a system's **signal-to-noise ratio**, which is simply stated as:

$$\frac{signal\ power}{noise\ power}$$

The power associated with the two signals can be found by calculating the signal and noise voltages' mean square values and dividing each by the circuit's output resistance. Thus, the signal-to-noise ratio is:

$$\frac{\overline{V_s^2}/R}{\overline{V_n^2}/R} = \frac{\overline{V_s^2}}{\overline{V_n^2}}$$

Where the line above the square voltages indicates the mean value.

Because the signal-to-noise ratio is a power ratio it is commonly expressed in decibels, where:

$$\text{signal to noise ratio (in dB)} = 10\log_{10}\frac{S}{N} = 10\log_{10}\frac{\overline{V_s^2}}{\overline{V_n^2}}$$

The signal-to-noise ratio required by a system to perform well is very much dependent on the system. A hi-fi system should have a signal-to-noise ratio of at least 70 dB or so in order that background noise is not heard in between music tracks of an LP, say. That of a telephone is not so important – about 40 dB is adequate. An excellent television picture will be possible with an aerial signal-to-noise ratio of 50 dB. Nevertheless, knowing a system's ratio allows a comparison with similar systems.

Noise figure

When a system comprises a number of parts (such as a transducer, connecting cable, interface circuits, etc.), and each part has its own signal-to-noise ratio, then some way is needed of calculating overall system signal-to-noise ratio. The way this is done is to give each part of the system a **noise figure**, sometimes called **noise factor**, where the noise figure, F, is given by:

$$F = \frac{\text{input signal-to-noise ratio}}{\text{output signal-to-noise ratio}}$$

and, because the noise figure, like signal-to-noise ratio, is a power ratio, it is also commonly given in decibels where:

$$F \text{ (dB)} = 10 \log_{10} \frac{\text{input signal-to-noise ratio}}{\text{output signal-to-noise ratio}}$$

However, as input and output signal-to-noise ratios are almost always given in decibels anyway, the noise figure (in decibels) can be calculated as:

$$F \text{ (dB)} = \text{input signal-to-noise ratio (dB)} - \text{output signal-to-noise ratio (dB)}$$

So, for example, a circuit with an input signal-to-noise ratio of 70 dB and an output signal-to-noise ratio of 65 dB has a noise figure of 5 dB. The lower the noise figure, the better the noise performance.

Once the noise figures of each part of the system are known the overall system signal-to-noise ratio can be calculated. This is done by first calculating the overall noise figure. In, say, the system of Figure 9.10, a transducer output signal is first amplified, then displayed on a moving-coil or similar display device. The reading obtained is to be a measurement of the measurand the transducer is monitoring.

Figure 9.10 *Using noise figures of each part in a system to calculate the overall signal-to-noise ratio*

The transducer has an output signal-to-noise ratio of 60 dB. Between the transducer and amplifier, and amplifier and display, are connecting leads with power gains (P_{G1} and P_{G3}) of 0.25, i.e., they are resistive and passive so

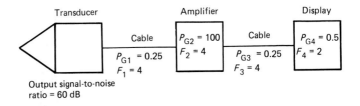

provide no gain – only attenuation. In such parts, noise figure (F_1 and F_3) is simply the reciprocal of the power gain: giving noise figures of 4. Similarly the display is passive, and with a power gain (P_{G4}) of 0.5, has a noise figure of 2. The amplifier has a power gain (P_{G2}) of 100, is an active part of the system, and merits its own noise figure of, say, 4.

Overall noise figure is calculated from the relationship:

$$F = F_1 + \frac{F_2 - 1}{P_{G1}} + \frac{F_3 - 1}{P_{G1} \cdot P_{G2}} + \frac{F_4 - 1}{P_{G1} \cdot P_{G2} \cdot P_{G3}}$$

This calculation can be extended to include any number of parts, as long as the power gains and noise figures of all parts are known.

In this example the relationship gives an overall noise figure of:

$$F = 4 + \frac{4 - 1}{0.25} + \frac{4 - 1}{100} + \frac{2 - 1}{25}$$
$$= 16.07 \text{ (about 12 dB)}$$

Overall signal-to-noise ratio is therefore the transducer's signal-to-noise ratio minus the interface system's noise figure, $60 - 12 = 48$ dB, which may or may not be sufficient, depending on the system.

The system of Figure 9.11 shows how the engineer can design a system to reduce the effects of noise. Here, the same system parts are used, except that a short length of connecting cable is used between transducer and amplifier – short enough to produce negligible power loss and so have a negligible low noise figure. This would be the case if a transducer with integral interface circuits was used to monitor the measurand.

Figure 9.11
Improving overall signal-to-noise ratio by locating the amplifier close to the transducer

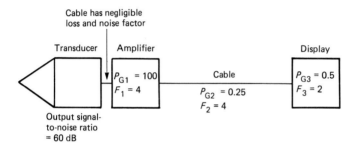

Cable has negligible
loss and noise factor

Transducer | Amplifier

$P_{G1} = 100$
$F_1 = 4$

Cable
$P_{G2} = 0.25$
$F_2 = 4$

Display

$P_{G3} = 0.5$
$F_3 = 2$

Output signal-
to-noise ratio
= 60 dB

Using the same relationship as before the new noise figure is now:

$$F = F_1 + \frac{F_2 - 1}{P_{G1}} + \frac{F_3 - 1}{P_{G1} \cdot P_{G2}}$$

$$= 4 + \frac{3}{100} + \frac{1}{25} = 4.07 \text{ (about 6 dB)}$$

So, the overall signal-to-noise ratio of the system is now improved by about 6 dB.

This result illustrates how it is vitally important to ensure the transducer is kept as near as possible to the first amplifying stage in the interface. Long connecting leads introduce noise which gives a low overall signal-to-noise ratio. Also, the illustration serves to show that the system's noise figure is almost totally dependent on the first amplifier's noise figure (because the noise figures of the other parts play a relatively small role in the overall noise figure calculation). It is for this reason that low noise preamplifiers are typically used in the first stage of transduction measurement interface circuits. Where possible, the low noise preamplifier should be integrally mounted in the transducer housing, also. This discussion has necessarily been a simplification of the matter of noise figure. Actual value of a noise figure depends to a large extent on factors such as temperature, frequency range, and the previous stage's output resistance. Saying that, for most purposes the calculations here are adequate.

Analogue and digital transduction systems

At some stage along the line, the engineer has to consider digital operation. As all transducers are inherently analogue in principle, this means that a conversion from analogue to digital is necessary. There are many techniques for doing this, touched on later.

The important thing about converting to digital operation is that any digital signal comprises a discrete value or fixed step within a range of values or steps. An analogue signal, on the other hand, can be *any* value within the range. So to convert accurately from analogue to digital means the engineer must ensure a sufficiently large number of digital values are available in the range.

The discrete values in the digital range are known as **quantization levels** and depend purely on the number of bits in the digital word used to define the value. For instance, if four-bit words are used to define the digital values, then only $2^4 - 1$, or 15 quantization levels exist. In general, 2^n levels are produced by analogue-to-digital conversion to an n-bit word.

This fact means that the resolution of a digital system is limited primarily by the quantization levels, because a particular digital word can only represent one particular analogue value. Analogue values in between the quantization levels cannot be accurately represented by a digital system. Increasing the number of bits in the words used by the digital system increases the system resolution, but the same level of resolution as an analogue system is theoretically never possible – no matter how many bits are used.

As far as conversion is concerned, we only need to consider here the conversion of the analogue transducer signal to a coded digital form acceptable to, say, a microprocessor-based measuring system. Such conversion is usually called **analogue-to-digital conversion** (ADC).

Most modern ADCs function on the principle that the analogue signal is turned roughly into a digital one, then the digital signal is turned back to an analogue signal (digital-to-analogue conversion – DAC – is inherently more straightforward) and compared with the original in a feedback process. If the original is larger than the reconverted signal the digital value is increased, if smaller the digital value is decreased, until the two are the same.

One of the most effective and useful ADCs is called the **successive approximation converter** (Figure 9.12(a)). In this, the signal from a comparator is applied to a register via control logic. At switch-on, the register is reset to zero. The DAC output is therefore zero and so any applied analogue input voltage causes the comparator output to be logic 1.

First the clock pulse causes the control logic block to change the MSB of the register to 1, and so the DAC output increases. If the analogue input voltage is still higher than this, the output of the comparator remains at logic one. At the next clock pulse the control block changes the next bit of the register to 1, and so on until each bit of the register is changed.

If the analogue input voltage after any of these *approximations* is lower than the DAC output, the control block resets the last bit to 0 before changing the next bit to 1. DAC output voltage in the form of a timing diagram is shown in Figure 9.12(b), for the four-bit successive approximation converter of Figure 9.12(a). Conversion time of any applied analogue signal to an n-bit converter is equal to n clock pulses.

One ADC becoming increasingly popular due to its

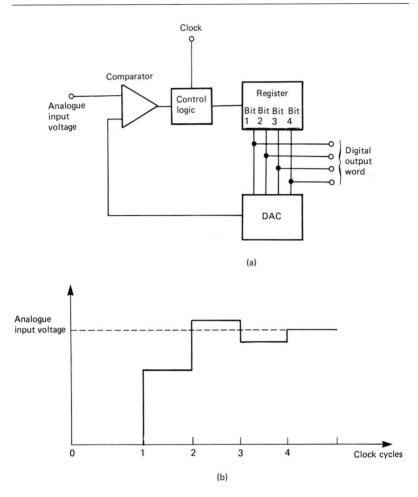

Figure 9.12
Successive approximation analogue-to-digital conversion: (a) block diagram; (b) timing diagram of DAC output

recent availability in easy-to-use integrated circuit form, and because of its high-speed conversion, is the **parallel converter**, sometimes called a **flash converter**, shown in Figure 9.13. The parallel converter uses a number of comparators to compare the analogue input voltage with a number of reference voltages. Each reference voltage corresponds to a quantization interval, so for a three-bit digital output (as shown) there must be seven available reference voltages (i.e., $2^3 - 1$) and hence seven comparators. The reference voltages are obtained with the use of a chain of resistors connected across the overall reference voltage: conversion accuracy is almost totally dependent on the accuracy of the reference voltages.

Conversion itself is simple and quick. The coder in the parallel converter uses the outputs of the comparators, converting them into the direct digital binary output required.

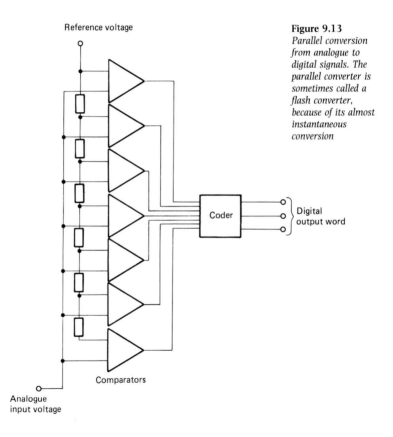

Figure 9.13
Parallel conversion from analogue to digital signals. The parallel converter is sometimes called a flash converter, because of its almost instantaneous conversion

Bibliography

Allocca, J.A., *Electronic Instrumentation*, Reston, 1983

Allocca, J.A., *Transducers, Theory and Application*, Reston, 1983

Bannister, B.R. and Whitehead, D.G., *Transducers and Interfacing*, Van Nostrand Reinhold, 1986

Brindley, Keith, *Modern Electronic Test Equipment*, Heinemann, 1986

Brindley, Keith, *Radio and Electronic Engineer's Pocket Book*, Heinemann, 1985

Cluley, J.C., *Transducers for Microprocessor Systems*, Macmillan, 1985

Norton, H.N., *Sensor and Analyzer Handbook*, Prentice Hall, 1982

Open University, *Instrumentation (T291)*, Open University Press, 1975

Pynn, C., *Strategies for Electronic Test*, McGraw-Hill, 1986

Seippel, R.G., *Transducers, Sensors and Detectors*, Prentice-Hall, 1983

Sydenham, Peter, *Transducers in Measurement and Control*, The University of New England Publishing Unit, Australia, 1975

Usher, M.J., *Sensors and Transducers*, Macmillan, 1985

Warring, R.H. and Gibilisco, Stan, *Fundamentals of Transducers*, Tab Books, 1985

Woolvet, G.A., *Transducers in Digital Sytems*, Peter Perigrinus, 1977

Index